WINGED CRUSADE

Winged Crusade

The Quest for American Air and Space Power

Edited by

Michael Robert Terry

Foreword by Walter J. Boyne

Imprint Publications
Chicago
2006

The editor and publisher dedicate their work in this volume to a beloved father, Robert Myron Terry (Lieutenant Colonel, Retired USAF, fighter pilot), and a beloved daughter, Jennifer C. Y. Cheung (RN), who passed away on December 27 and 28, 2005, respectively.

Cover design and illustration by Chris Hureau

Library of Congress Control Number 2006933977
ISBN-13: 978-1-879176-43-0
ISBN-10: 1-879176-43-2

Military History Symposium Series of the United States Air Force Academy, Vol. 9
Mark K. Wells, Series Editor

Made in the United States of America and printed on acid-free paper

To those who love the vastness of the sky

Military History Symposium Series

of the

United States Air Force Academy

Vols. 1-6 (Carl W. Reddel, *Series Editor*)

An American Dilemma: Vietnam, 1964–1973 (1993)
Edited by Dennis E. Showalter and John G. Albert

A Revolutionary War: Korea and the Transformation of the Postwar World (1993)
Edited by William J. Williams

The Intelligence Revolution and Modern Warfare (1995)
Edited by James E. Dillard and Walter T. Hitchcock

Tooling for War: Military Transformation in the Industrial Age (1996)
Edited by Stephen D. Chiabotti

Forging the Sword: Selecting, Educating, and Training Cadets and Junior Officers in the Modern World (1998)
Edited by Elliott V. Converse III

Air Power: Promise and Reality (2000)
Edited by Mark K. Wells

Mark K. Wells, *Series Editor*

Future Wars: Coalition Operations in Global Strategy (2002)
Edited by Dennis E. Showalter

Prisoners of War: The American Experience (2004)
Edited by Vance R. Skarstedt

For further information on this series, visit the website at
www.imprint-chicago.com

Contents

Preface

Picking a title for an academic conference and the resulting volume of varied scholarly essays is not an easy task. Our goal for the 20th Military History Symposium was to capture the power and excitement of the Wright brothers' monumental effort to fly. One of the principal meanings of the word "crusade" is, "a vigorous concentrated action for some cause or idea." Under this definition, the Wrights' near-decade-long efforts fit perfectly. Their single-minded determination to succeed demonstrated the very best of all of humanity: courage, determination, imagination, and curiosity. Beyond overcoming the daunting aeronautical engineering challenges, they also inspired future generations of airmen and established a laudable vision of air and space power which stretches from the sandy soil of North Carolina's outer bank to the planets and stars and beyond.

We deliberately framed our examination of this vision around the American experience for a variety of reasons. First, of course, is that the Wright brothers exemplified the pioneering spirit of Americans in the late nineteenth and early twentieth centuries. Although they most certainly capitalized on the knowledge and experiences of scores of their air-minded predecessors from around the world, the Wright brothers were the ones who finally solved the myriad of complex challenges leading to powered, controlled, and sustained flight. Much of what followed over the decades featured Americans in central roles. Even if advances in aeronautics took place elsewhere and featured incredible creativity by those in other countries—as they most certainly did—the contrails of this "Winged Crusade" inevitably seem to track back to the extraordinary vision of Americans. America was not just the world's first aerospace nation, but it also remains in its vanguard. And, finally, as the essays in this impressive volume document, many of these American air power visionaries were scientists and engineers as well as military professionals and civilians. It was important to comment on the contributions of this diverse group.

This volume marks only the third time in the nearly forty-year history of the Military History Symposium that we have focused principally on the subject of air power. As ever, we are deeply grateful to the Air Force Academy's Association of Graduates for their unwavering support, both for the symposium and resulting volumes. The impressive group of scholars who have contributed essays here enjoy substantial national and international reputations in the field. Their research and conclusions reflect the most contemporary—I dare say "leading-edge" thinking—in air and space power history. Each contributor is also accustomed to the controversy which occasionally arises by any discussion of air power and its powerful impact on warfare over time. It is virtually impossible to pick up any serious journal on military or political affairs nowadays and not be subjected to the wisdom of commentators on air power and its application in contemporary conflict. Sadly, too often these analyses are carried out by the profoundly misinformed or those with obvious agendas.

Shallow arguments or relentless campaigns against air power do little to educate future military leaders or civilian policymakers about the usefulness of this vital component of a joint security structure. If, by studying this volume, a reader comes away with nothing else, it should be the reassurance that American airmen fully understand the role, impact, abilities, and limitations of air power in all kinds of warfare. More importantly, their quest for understanding and improvement of air power for the defense of this nation continues unabated. Surely this qualifies as a "Winged Crusade" in the most uplifting and positive way.

Mark K. Wells
Series Editor

Foreword

It is both a privilege and a pleasure to write a foreword to this collection of essays from some of the most able air and space historians of our time. While it would be impossible to capture, in a foreword, the essence of each of the essays, it is possible to point out their connections, and how they complement each other. Further, these connections prove there are often far fewer than six degrees of separation when it comes to the great leaders and significant events in air power.

For with aviation comes passion. One cannot love the idea of flight abstractly; it takes hold of one's soul and permeates the conscience. Trying to figure out why it does, one comes to a belief that it is somehow metaphorically genetic. It must be that passion about flight flows through one's veins, permeates one's genes, and saturates the collective consciousness as breaking the bonds of earth has captured the imagination of the world for millennia. The Wright brothers and all the other pioneers of flight were afflicted with the same passion, wonder, and excitement that continue to motivate the Winged Crusade today. The identity with the ideas of the Winged Crusade has extended down through the last century and into the present like wild fire. American aviation grew with bigger, faster planes, jets, missiles, space defense, and most of all with men and women willing at every point to risk their lives for their passion.

The passion leads to not being quite normal in the conventional sense. Normal people do not come to the hard work, the self-denial, the consistent effort it takes to succeed. To translate the passion of flight into a potential unprecedented power took dedication and sacrifice by those visionaries and aviation professionals.

Perhaps the greatest connection for historians—and almost certainly their greatest inspiration—are identifying and honoring the sacrifices the Wright brothers' progeny, generations of American aviation professionals, have made from the dawn of U.S. military aviation until this very moment.

The concept of service before self began with the tragic 1908 death of Lieutenant Thomas Selfridge, continued through peace and war for the next ninety-five years and will go on for as long as there is a United States and a United States Air Force to defend it. And no one is more conscious of this climate of selfless giving than the historians who investigate the Winged Crusade's numberless acts of heroism over the years. Sacrifice has indeed been the very soul of the Winged Crusade, first in terms of their lives, some 54,000 having made the supreme sacrifice, but also in terms of career choices and setting family priorities.

It is the historian's welcome duty to record those acts to highlight just how strong the predecessors of the Winged Crusade exemplified courage and moral right. The United States is blessed today with an extraordinary number of air power students, men and women whose experience ranges from the upper levels of academia to the down-and-dirty world of combat airmen.

It was not always so. For too many years advocates of air power tended to be biased, and indeed, too often strident, even when they did not have personal experience in the exercise of air power. This was doubtless necessary and even beneficial during the years when air power attempted to assert itself. Struggling to gain independence, advocates sought to acquire and exercise the technology that would lead to a war-winning capability. Looking back, we have to understand that these eager Winged Crusade pioneers knew that the technology their inflated claims required would in fact come into being in time. Most realized the new, empowering, technology they sought would have to be combined with training and doctrine to create a war-winning instrument. However, to convince others, especially the many hard-core opponents of air power, they pushed the envelope of belief, claiming capabilities not yet in hand.

Modern advocates have seen technology, training, and doctrine of air power advance at an extraordinary pace. Its demonstration in the Gulf War, in the Balkans, and in Operation Iraqi Freedom exceeded not only the claims, but also the wildest dreams of their predecessors.

Yet the advance in air power over the century was not an even process, and it was, oddly enough, compromised in 1945 by one of the most extraordinary developments in technology, the introduction of nuclear weapons. These new weapons offered a convincing basis for "victory through air power" as long as the United States maintained a nuclear monopoly. However, once the Soviet Union broke the monopoly, the United States viewed nuclear weapons not as the means of winning a war, but rather as the means to deter a mutual catastrophe.

In some respects, this development moved the concept of air power as a war-winning weapon back to the proverbial square one. A number of other factors compounded the difficulty. First was the towering size and strength of the Soviet Union's conventional forces. The United States was incapable of matching the Soviet Union in numbers of men, tanks, artillery, or aircraft. Second, the cost of weaponry rose dramatically at a time when the United States faced budget restrictions competing with increasing demands for critically important domestic programs.

It was of course essential to maintain the concept of "mutual assured destruction" so no cataclysmic nuclear exchange took place, while engaging the Soviet Union and their proxies around the world. These limited wars drained American resources, and sadly, the national will. Faced with the numbers/cost dilemma, the only path to success was through technology, a path that followed so successfully that the United States—for the time being—is now the only true superpower, with capabilities based not on nuclear destruction, but on air, space, and information dominance.

Each article of *Winged Crusade* touches upon this theme in one manner or another, and in so doing defines the modern air power advocate. The papers are instructive because they reflect the authors' patient assessment of the past in drawing logical predictions about the future. Unlike their more jingoistic predecessors, these students of air power look as closely at failures and at enemy capabilities in relation to American successes and technological triumphs.

In his fine essay, "Wright Brothers: Pattern for American Air and Space Power—Innovation and Legacy of Ingenuity," Tom Crouch points out it is more necessary than ever to examine the American experience with air power a century after the Wright brothers' first flight. While suggesting there is something inveterate in the American national character that especially suits Americans to lead in aviation, he also states that our current air power mastery automatically incurs bitter global opposition and competition. The United States' high-tech Air Force is aging, and as Crouch notes, our terrorist opponents have introduced new challenges, by inverting our technology and using it against us. The question now arises as to whether or not the United States can display the same resilience in our response to terror as we did to the nuclear stand-off with the Soviet Union.

The terrorist movement differs from previous threats to our security, most notably in the fact that we are confronted with a movement not controlled by a national government. (At least, not openly controlled!) The unthinkable might occur at any moment with the detonation of a weapon of mass destruction in one or several of our major cities. Yet whether this detonation occurs tomorrow or ten years from now, it will not affect the long-term ambitions of the terrorists, who assure us they are planning for the centuries, not the years nor even the decades.

We can obtain some insight into the hazards of combating of such a long-term threat from Tami Davis Biddle's paper, "An American Way of War: The Quest for High-Altitude Daylight Strategic Bombing." As Biddle points out, the United States spent many decades pursuing the quest to perfect "precision bombing," unattainable for years until the Air Force developed the appropriate bombing equipment and suitable ordnance. By then, American airmen found the enemy had changed, no longer presenting targets against which the power of high-altitude precision bombing could be projected effectively. In our current battle, the terrorists will not provide us the option of a decades-long quest for a solution. We will have to obtain that solution in the near term if we are to survive.

I may be grasping for straws, but I take comfort in a parallel suggestion drawn from James Corum's paper, "*Luftwaffe* Intelligence: How it Viewed the United States Army Air Forces." The *Luftwaffe*'s intelligence efforts on the U.S. Army Air Forces, Corum notes, reflected the general performance of the German military as a whole. At operational and tactical levels, the German intelligence efforts could be very good, but the failure—fortunately for the United States—lay at the top of the *Luftwaffe*'s leadership, and in Adolf Hitler's inner circle. Their complacent, ill-informed, indeed, provincial, view characterized the United States as a decadent democracy incapable of producing the necessary quantities of high-grade aircraft and pilots. It is to be hoped that a similar arrogance holds true with the leadership of the terrorist opposition in Iran, al Qaeda, and elsewhere and dooms their efforts to failure. While they certainly cannot have any illusions about American technical ability, they may be as much in error about the American public's inability to endure further terrorist attacks as the Nazi High Command's estimate regarding American industrial might.

This is not the first time we have faced a hardy, difficult-to-engage adversary, and as Merle Pribbenow points out in "Rolling Thunder and Linebacker Campaigns: The North Vietnamese View," we may find that our interpretation of "victory" in the war on terror will not correspond to the interpretation of the terrorists. In an insightful analysis of Linebacker II, Pribbenow makes it clear that while both sides may claim a military victory in a long and hard-fought battle, it is far more difficult to determine who won the political victory. In the case of Linebacker II, most U.S. air power advocates contend the outcome of the battle—the destruction of the North Vietnamese integrated air defense system—brought the North Vietnamese politicians back to the negotiating table in Paris, and made the 27 January 1973 Armistice possible. However, the U.S. Congress had already guaranteed the political victory to the North Vietnamese by signaling its intention to cut off appropriations to prosecute the war. The parallels here between the Vietnam War and the current conflict in Iraq have never been more obvious, nor more ominous.

It might be wise now to turn to Herman Wolk's thought-provoking essay, "Who's in Control? A Century of Organizing for Air War," to refresh our minds about the definition of air power. Wolk provides two conventional definitions of air power, the first being that air power is "power adapted to military ends, the power which a nation may use in war against another nation or nations, based upon a national weapons system organized about air forces." The second definition is from the revered General Henry H. "Hap" Arnold, a true architect of twentieth-century air power, who defined it as "a nation's ability to deliver cargo, people, destructive missiles, and war-making potential through the air to a desired destination to accomplish a desired purpose." Wolk goes on to say that it is the evolution not only in technology, but also in organization and control of air forces that enabled American air power to become dominant. He further notes that American air power is now uncontested, but makes the key point that the advances in organizing and controlling air war were made by men of faith who believed that evolving technology demanded control by airmen.

This last point is re-echoed in an essay, which emphasizes the importance of the airman to air power. In "Forging American Air Power: Patrick, Arnold, and Doolittle," Dik Daso reviews the efforts of three men who were fundamentally important to air power in very different ways, but who today, sadly, are virtually unknown outside the realm of air power historians. Most underrated of all, Major General Mason Patrick made the U.S. Army Air Service work, Army fashion, while allowing it to advance as far as it could on the limited budgets of the time. In Daso's words, he "legitimized the Air Service," playing a vital transitional role in its evolution during a time of extraordinarily tight budgets. Daso also selects Lieutenant General James H. "Jimmy" Doolittle for inclusion. Doolittle was a great warrior, most famous for leading the 1942 raid on Tokyo and for his later command assignments in Europe. The public generally recognizes his racing feats, but tends to forget he was also a genuine scientist, whose persistent advocacy for the development of 100-octane gasoline before World War II had a tremendous, and largely unacknowledged effect. Somewhat surprisingly, considering the relative amount of time and effort he expended in achieving it compared to his later endeavors, Doolittle believed his work in instrument flight to be the most important contribution of his career to aviation.

Daso's third figure, "Hap" Arnold, while not a scientist, was prescient in advocating experimental research. The importance of this viewpoint has to be understood in the context of the times, when there was scarcely enough budget for the salaries of the personnel, much less the acquisition of equipment. It was the beginning of an Arnold approach that would have signal success, the appeal to American universities and to American industries to shoulder the initial burden of aeronautical research, rather than to depend upon military funding. As Daso notes, Arnold thus became the author of the Military/Industrial/Academic (M/I/A) complex, a much healthier idea than the label attached to the "military-industrial complex" of the post-Eisenhower era.

Daso's paper ties in neatly with those of Thomas Keaney and Jacob Neufeld. Keaney, in his "Jet Aircraft and Defense Imperatives, World War II to Vietnam," notes that General Arnold sought to further his ideas on the importance of research by asking Dr. Theodore von Karman, a prominent scientist in the field of aeronautics, to lead a special study. Von Karman, working with thirty-six leading scientists, produced an epic work, *Toward New Horizons*. Von Karman summarized the work in the first volume, *Where We Stand*. In it, he charts the course of future air warfare to include supersonic flight, intercontinental missiles, vastly improved communications and, of course, jet and rocket propulsion. And from this first formal assemblage of the M/I/A congress came first the Scientific Advisory Group and later the Scientific Advisory Board, which Doolittle would chair from 1955 to 1958.

Neufeld, in "General Bernard Schriever: 'Father of Air Force Missiles and Space Power,'" again touches on Arnold's uncanny ability to pick personnel who could deal with the eternal mysteries of future air and space power. Arnold chose then Colonel Bernard Schriever to head the Air Staff's Scientific Liaison Office to ensure continued success in the partnership between airmen and scientists. If ever there was "the right man for the right job at the right time," it was his choice of Schriever. His brilliant intelligence and management skills would create for the United States a vital succession of intercontinental ballistic missiles, and simultaneously prepare it for its future domination of space. Neufeld quotes Schriever as saying, "our safety as a nation may depend upon our achieving 'space superiority.' Several decades from now the important battles may not be sea battles or air battles, but space battles, and we should be spending a certain fraction of our national resources to insure that we do not lag in obtaining space supremacy. Besides the direct military importance of space, our prestige as world leaders might well dictate that we undertake lunar expeditions and even interplanetary flight." Not even the notably irascible and hard to satisfy Arnold could have been discontent with such a universal view, one that has served our country so well.

Schriever's grasp of technology, Arnold's thirst for research, and that mysterious quality inveterate in the American psyche all seem to be combined in the group's fascination for aviation and especially, its exercise in warfare. John Guilmartin provides insight into this in "The Aircraft That Decided World War II: Aeronautical Engineering and Grand Strategy, 1933–1945, the American Dimension." His viewpoint is especially important today when unthinking or ill-motivated critics claim that air power has not been proven in Iraq, or more recently, in Israel's operations in Lebanon. Air power has already been proven, Guilmartin clearly points out, in World War II, when three air campaigns were strategically decisive—the Battle of Britain, the Combined Bomber Offensive, and the strategic bombing campaign of Japan. And in this regard, aircraft, always the first love of the air power advocate, are critically important. Guilmartin stirs our emotions with his swift characterization of the great aircraft of the period, but notes while strategically pivotal to the war effort, aircraft design was dependent upon far less well-known contributions for its success. Among them, Guilmartin singles out Jimmy Doolittle's 100-octane fuel, and the turbo supercharger development by Dr. Sanford A. Moss and perfected by General Electric. Guilmartin's further comments on the reasons for America's air power success in World War II is in fact a cause for alarm. While much of the American success was due to the "sheer depth, breadth, and vitality of America's aviation industry," he goes on to note, aviation technology does not have the same strategic primacy today as it had in the interwar period. Further, we have to be aware that today's world of declining budgets, smaller production runs, fewer new aircraft, longer development times, and a decline in public interest pose a threat to the nation's security.

In closing, these papers are to be savored and used, for they contain a remarkable amount of information, and an even more remarkable amount of insight. They are immensely valuable for their cogent views of the past, for from them may be drawn important ideas on the future. And I think it is remarkable that nine papers, by nine scholars, on nine very different subjects, should be so internally consistent, and often, so very closely connected.

Walter J. Boyne

Introduction: Winged Crusade

Michael Robert Terry and Vance R. Skarstedt

As this ninth volume of the United States Air Force Academy's Military History Symposium Series was about to go to press, a grateful America dedicated a special memorial to its airmen in Washington D.C. It also began a year-long celebration of the 60th anniversary of the Air Force's creation. As President George W. Bush pointed out at the dedication ceremony, "air power is still a relatively new phenomenon" and "there are still Americans alive today who were born before man had ever flown." Despite its newness, the President stated, "it's hard to imagine a world without the Air Force protecting us in the skies above." In just a short time, air power grew to incorporate space power and progressed from being the dream of two brilliant brothers to become what the President called an indispensable tool in our nation's arsenal.[1]

This edition of the Academy's symposium series commemorates the centennial of flight initiated by those two brothers, Orville and Wilbur Wright. This volume's title is *Winged Crusade,* and in this context, the term "crusade" symbolizes the Wright brothers' quest for sustained, controlled, and powered flight and how their success began. They demonstrated remarkable curiosity, intuition, imagination, and undaunted courage in their quest. They connected the American pioneering spirit to humankind's desire for flight, and captured the imagination of the world. Their accomplishments led to new ways of thinking about transportation, commerce, and inevitably, warfare. These proceedings not only discuss the journey of two remarkable self-educated engineers, but also their legacy. Theirs is a legacy that exists in a world made smaller by air travel, in man's venturing into space, and most importantly, in the military air and space power of the United States and the world.

Orville and Wilbur Wright achieved success through discovery. They determined that the control of flight involved more than just achieving stability. It required positive control in three dimensions. Their demonstration of controlled, as well as powered flight, gave birth to what eventually became the central pillar of air power.

Their discovery went beyond kite-flying, glider trials, and wind tunnel experiments. By "visualizing abstractions" they solved the inherent problems of instability and paved the way for controlled flight. Their approach evolved from amateur observations to trial-and-error experiments, and then to practical engineering techniques and processes.

Essentially, they combined engineering and experimentation to become the first test pilots. On his successful first flight, Orville became aware that flying requires constant adjustment of the control surfaces to the prevailing flight conditions. They established the principle that control was not merely an adjunct to stability, but its very essence. Their insight and methodology established the innovative pattern that

1

characterized the development of aviation in America. By inventing the first practical aircraft, they became the founding fathers of air power. Their names symbolize the first century of the "Winged Crusade"—a century of remarkable progress that includes intercontinental flight, breaking the sound barrier, and landing on the moon.

The Wrights' successors, generations of American military airmen, used the same spirit of discovery and innovation. Often challenging accepted limitations, the men and women of the Winged Crusade envisioned a dynamic future for air and eventually space power in the nation's defense. By defining doctrine, technology, organization, and independence within the national defense structure, military airmen forged a "flight path" for air power's growth. In only one generation, air power went from artillery-spotting to high-altitude, strategic precision bombing. Within another generation, the United States, as well as most other nations of the world, established its own independent air force providing intercontinental military options.

Expanding on the airman's perspective originated by the Wrights, American airmen navigated the dynamic and turbulent early decades of air power. It is fair to say that air and space power has never been completely defined nor its potential totally realized. In this regard, the quest and the crusade continue.

The articles in this volume, written by nine leading air and space historians, provide a comprehensive view of the history of American air power. Thomas Crouch, biographer of the Wright brothers, details their innovative process. His article provides a useful framework for the remainder of the articles. Herman Wolk examines how national air power depended on the search for an institutional identity in order to provide it with an organizational structure to implement national objectives. Dik Daso discusses the contributions of three important pioneering airmen—Generals Mason Patrick, Henry H. "Hap" Arnold, and James Doolittle—who laid the groundwork for the eventual creation of the United States Air Force. Tami Davis Biddle traces the development of precision-bombing doctrine, a promise unattainable for decades. James Corum's article analyzes the role of intelligence in the projection of air power during World War II, and shows that an enemy deals with Clauswitzian friction and the limitations of intelligence in their own way. John Guilmartin presents a new perspective on assessing the "strategic" measurement of air power, not necessarily based solely on technological aircraft design. Merle Pribbenow evaluates the Vietnam air war using a "cultural-institutional" viewpoint, and points out the fine line between failure and success in aerial warfare. Thomas Keaney highlights the post–World War II "technological revolution" in aircraft design. Jacob Neufeld reviews Bernard Schriever's vision to extend air power into space.

Tom Crouch's essay opens this volume in a powerful manner. He surveys the role air power played in developing an America that commands the skies and leads the world in space. He frames America's achievement by pointing to our national psyche as personified by Wilbur and Orville Wright: Quintessential middle [class] Americans . . . exemplifying our national strengths and values made good through hard work, commonsense, perseverance, and native ingenuity. Describing the critical thinking of the Wright brothers, Crouch examines in detail the development of the Wright *Flyer* from 1899 to 1903. He articulates their design efforts by describing the five aircraft they built in four years. They blended scientific inquiry, engineering insight, technology, craftsmanship, and test-pilot courage to succeed.

Almost immediately, other aviation pioneers came forward. Glenn Curtiss, Alberto Santos-Dumont, and Henri Farman soon flew farther and longer than the Wrights; Louis Blèriot flew across the English Channel in July 1909 demonstrating the potential impact of aviation to international boundaries and national security. While all of these accomplishments are significant, Crouch argues that none of those individuals would have flown had it not been for the Wrights. The history of aviation, according to Crouch, begins with the Wright brothers. Competitive races, richer prizes, and the threat of war not only generated technological changes, but also cost America its lead in aviation albeit temporarily. By the time of Orville Wright's death in 1948, the United States reassumed its lead in aviation, largely due to the technological advancements associated with the two world wars.

Herman Wolk's article examines the political aspects of the Winged Crusade. He argues that national air power depends on its organization and employment. This, of course, is not solely a technological issue. It is bound up in domestic politics, international threats, economic realities, service rivalries, and even personality conflicts. Wolk leads the reader through the political obstacles faced by air-minded army officers during their attempts to expand the role of air power and traces their progress in relation to the significant issues of World War II.

The National Security Act of 1947 created an independent Air Force and with it an organization commanded by experienced airmen who knew how to organize and operate air forces. In its first three decades, the United States Air Force met the conventional challenges of the Korean and Vietnam Wars and successfully adapted its organizational model to face new regional threats and the responsibility of nuclear deterrence. With the end of the Cold War, the United States faced a more conventionally armed potential threat so the Air Force streamlined its organization by merging the tactical and strategic commands. The Air Force further organized into ten Aerospace Expeditionary Forces to hone its theater-specific war-fighting, and support of joint operations.

Wolk concludes that American air power is now uncontested, but also echoes Arnold's warning of almost fifty years ago that air power itself can become obsolete. Since the implosion of the Soviet Union, a new threat has arrived. This new menace involves a transnational terrorist organization potentially armed with weapons of mass destruction. As Wolk adroitly points out, the need to determine how best to organize the American air and space arm to meet today's challenges and those in the future continues.

Dik Daso's article discusses three men who came to personify the Winged Crusade. The history of American air power is fortunate that Patrick, Arnold, and Doolittle lived and served when they did. These pioneers, while often unappreciated, made contributions that proved to be essential in the development of American air power. Mason Patrick, although becoming an aviator late in life, assumed a pivotal role in the interwar years as chief of the Army Air Service. By sending the often controversial Billy Mitchell off on world tours to study other nations' efforts in air power, he removed a sore spot for Army leadership and began to serve as a mediator between the tradition-bound Army and the often zealous officers of the Army Air Service. Working within the system, Patrick made possible the progress of many technological and doctrinal air power advancements.

Throughout his career, Henry H. "Hap" Arnold used imagination to solve the technological and logistical challenges involved with projecting air power anywhere in the world. His notions of research and development augmented by the practical experience of deploying remain the backbone of expeditionary air forces today. Arnold pursued the concept of "aerial refueling" and led deployments, such as the July 1934 exercise relocating aircraft and personnel from Wright Field, Ohio to Fairbanks, Alaska. This proved an important demonstration of the potential for global air operations. He, more than anybody, helped give the Army Air Corps its "wings."

James Doolittle exhibited "guts," both physically and morally, as a test pilot examining aircrew physiology and airframe structural integrity. In exploring the extreme limits of human endurance capacity and aeronautical engineering, Doolittle researched and flight-tested performance, navigational, and attitudinal instruments "under the hood" to create the capability of "blind-flying" so airmen could descend through adverse weather phenomena to land safely. Most Americans know how Doolittle's leadership and savvy enabled the United States to bomb the home islands of Japan only four months after their attack on Pearl Harbor. It was his work in aviation research and development before the war, however, that proved to be his greatest contribution. Intestinal fortitude and moral clarity to make the tough decisions, Daso points out, shaped Doolittle's character in persevering for the expansion of American air power and for the crucible of combat as a leader and commander. Doolittle had "guts."

Tami Davis Biddle's "An American Way of War: The Quest for High-Altitude Daylight Strategic Bombing" analyzes the evolution of American doctrine for precision bombing from its origins in World War I to Operation Iraqi Freedom. Her detailed study discusses how the theories espoused by the airmen of the early years of military aviation were thwarted by the limited technical capabilities of the day.

According to Biddle, American airmen possessed a unique air vocabulary. They developed a language of terms used to describe bombing techniques and procedures. They spoke in a vernacular that indicated to the uninitiated that sophisticated expertise comprised their theories. As Biddle makes clear though, the real acceptance of their identity by rival services and national leadership had to be proven in war and this, more than anything else, led to the establishment of a successful doctrine. Every strategic bomber lost over Germany distracted from the process; every aircrew shot down undermined the rhetorical promise that direct attacks upon an industrial nation could be a war-winning strategy. However, adherence, with some tactical modifications to the original theories of airmen espoused between the world wars, won out. The United States Strategic Bombing Survey concluded that "Allied air power was decisive in the war in Western Europe." This revelation, according to Biddle justified the perseverance and sacrifice. The atomic bombs dropped on Hiroshima and Nagasaki, and the contribution air power made toward eroding German industry and preparing the way for the invasion of Europe in 1944 shaped the postwar national defense organization and ultimately established an independent Air Force on the eve of the Cold War. The Winged Crusade was now in control of its flight path.

The Korean and Vietnam Wars saw mixed results in the use of air power. Despite the success in preventing nuclear confrontation with the former Soviet Union, America's Air Force faced new challenges in what became known as "limited war." Scores of commentators and "experts" pointed to air power's failures with little understanding how it had been grossly misapplied. Nevertheless, almost two decades later, air power contributed to the most lopsided defeat in history against Saddam Hussein's Iraq in the first Persian Gulf War. As Biddle points out, the promise of precision air power finally satisfied the promises of the lectures resonating from the Air Corps Tactical School. Yet, as Biddle concludes, the ability of the United States Air Force to hit any target in the world with accuracy—the essence of precision bombing—does not mean that wars involving U.S. forces will become simple with predictable successes. For many, no amount of empirical evidence is sufficient. Air power must always prove its "airworthiness."

That air power and military intelligence are inextricably linked has long been known. James Corum's article reminds us that the enemy—a willful, thinking entity—wages war simultaneously and will attempt to undermine plans, deceive minds, defend its vital interests, and defeat others' will to fight. He examines the German view and understanding of the American air effort during World War II, and provides an organizational analysis of the *Luftwaffe* intelligence. Perhaps, more importantly, Corum's article provides critical insights into the relationship between the nature of intelligence and *Luftwaffe* operations. He describes how the *Luftwaffe* effectively used and misused the intelligence on American air power, and assesses the general effectiveness of the *Luftwaffe* intelligence system.

Despite having an effective intelligence collection network, the Third Reich did not have a central point to comprehensively evaluate the vast quantity of data. According to Corum, even if the Germans had been better able to analyze strategic intelligence, Adolf Hitler had little use for strategic analysis that contradicted his own genius. America simply did not figure into the Führer's grand strategy. Therefore, Hitler did not believe German intelligence when they predicted the United States could satisfy and even exceed Franklin Roosevelt's call for 50,000 planes. Of course, such underestimation and contempt by senior German leaders toward the United States led to many gross miscalculations. By late 1942, the *Luftwaffe* revised upward its pilot training program needed to defend the Reich from the Americans. By then it was too late. As Corum concludes, when the air war turned to attrition, the inability of German pilots to survive in combat against better-equipped and better-trained Americans became the crucial factor in losing control of Germany's airspace. Ignored and often misrepresented intelligence confounded the *Luftwaffe*.

While doctrine and intelligence play equally important roles in the use of air power, the technology of air power is fundamental to its success. John Guilmartin's "The Aircraft That Decided World War II" addresses this. Guilmartin argues that the air war of World War II comprised a global conflict in ways that the war on land and sea did not. The success of air power in World War II was *not* the result of superior aeronautical design, but how those designs influenced the prosecution of the war. In Guilmartin's words, "It is easy to conclude that strategic advantage obtained in the air must have flowed from superior design and that is not always the case."

Guilmartin relates successful aircraft design to combat employment effect. He articulates how aircraft performance influences tactical effectiveness. Economic costs as portrayed in terms such as reliability, maintainability, and in-commission rates impact air power employment. Handling characteristics affect operational wastage in that aircraft are less likely to be lost to pilot error than enemy activity. Strategic circumstances are often only one consideration in the design of combat aircraft. In addition to providing important details regarding aircraft flown by American, British, German, and Soviet fliers, Guilmartin analyzes the same relationship of production, performance, operational reliability, and flying qualities to tactics, operations, and strategic thinking. He clarifies some important, perhaps timeless relationships between technical aero designs, tactical employment, and strategic vision.

Merle Pribbenow's contribution to the volume provides some of the most riveting comparisons between American air power and the North Vietnamese Air Defense Command (NVADC) air defense measures. His article reveals some of the flaws of the NVADC's air defense doctrine and outlines their strategy for defending their airspace. The NVADC did not underestimate the threat posed by U.S. air power. The combat between American air and North Vietnamese air defense proved deadly and often saw the advantage swing back and forth. This air war became one of action and reaction by both sides in an attempt to keep the advantage. The 1964 U.S. air strikes after the Gulf of Tonkin incident demonstrated a number of shortcomings in North Vietnamese air defense. The North Vietnamese, with help from China and the USSR, quickly upgraded their organization, air defense technology, and command. The United States also adapted to the changing combat environment over North Vietnam. American fliers were forced to adjust to the "political war" of Vietnam with its restrictive rules of engagement, and the North Vietnamese had to adapt to the challenges of a technical war.

Pribbenow suggests there were many occasions during the first few years of the war where a sustained and unrestricted application of American air power might have led to a different outcome. In his view, subsequent American analysis that the war was lost from the beginning may be a false premise. The North Vietnamese view of the Rolling Thunder campaign brings to light that this air offensive had a more damaging impact on North Vietnam and came closer to affecting a different outcome than previously thought.

Pribbenow's view from the other side of the fence provides an important perspective of adversaries waging aerial warfare. His insight provides a better understanding of the dynamics of air warfare and delineates more clearly the line between the failure and success of air power.

Thomas Keaney's essay, "Jet Aircraft and Defense Imperatives, World War II to Vietnam," addresses the technological revolution and fundamental realities that confronted the Air Force at the dawn of the jet age and ushered in a new generation of combat aircraft. Keaney frames the challenges of technological design for postwar aircraft around economic constraints, strategic challenges, perceived threats, and institutional roles and missions. Jet propulsion and swept-wing design technologies as well as the potential challenges from the Soviet Union stimulated rapid developments within the American aircraft industry. The Cold War required expediency in the American pursuit of long-range bombers capable of delivering nuclear weapons against the Soviet Union. Keaney analyzes such factors as air refueling, fighters for additional protection, interim measures to extend longevity of existing bombers, strategic and operational considerations for deployment, and the impact of technical and competitive design considerations. His article analyzes many of the technological considerations for designing aircraft in the context of a possible nuclear war with the Soviet Union and the conventional means of the Korean War. The strategic and conventional requirements of Cold War air power determined the technical developments of American fighter aircraft. Keaney traces the course of jet fighter development between multi-role emphasis to specialized roles and back again through out the 1950s, 1960s, and 1970s.

The Winged Crusade also began a transformation from air power to air and space power in the late 1950s as the domain of space became a major issue. Jacob Neufeld, writing in "General Bernard Schriever: 'Father of Air Force Missiles and Space Power,'" demonstrates how the Air Force made this transformation. In a February 1957 speech entitled "ICBM—A Step toward Space Conquest," General Schriever warned of the vital necessity for the country to achieve space superiority, not only in preparing for eventual space battles, but also for prestige as a world leader. Schriever's call to one day land on the Moon was echoed by President John F. Kennedy. Schriever's call for space weaponry would be repeated a quarter-century later in President Ronald Reagan's 1983 Strategic Defense Initiative. Responsible for nuclear deterrence, the Air Force eventually used satellites and space to integrate the ICBM as an effective way to protect the nation. Schriever continued to promote the Air Force's commitment to space superiority, arguing that having superiority in space is a fundamental requirement since "On it will depend the entire future position, prestige, and indeed, the welfare of the United States."

With several significant space milestones achieved by Schriever's ICBM team, the Air Force organized for space. Schriever led the space research and development programs. The Air Force assumed responsibility for reconnaissance and surveillance satellites and for developing and launching all military space boosters. According to Neufeld, Schriever continued to press for exploiting the advantage that space offered to our "vital military deterrent posture." True realization of space's importance came in 1959 after two editions of "Basic Air Doctrine," first published in 1953. The revised manual established a new horizon in Air Force thinking. The transformation was one of doctrinal scope, as illustrated by the change in the manual subtitle from "air power doctrine" to "aerospace doctrine."

Schriever also advocated unmanned space vehicles such as "surveillance and reconnaissance, interception, bombardment, and command, control, and communications" for Air Force space roles. He also promoted manned space vehicles such as Dyna-Soar (X-20) and the Manned Orbiting Laboratory. Despite cancellation of both military manned space programs, the Air Force enjoyed great success with respect to boosters and instrumented satellites for communications, weather, reconnaissance, warning, and navigation. Schriever's legacy is epitomized by the Air Force's leading role in America's space effort and the activation of the Air Force Space Command in September 1982. Jacob Neufeld's piece is testament to Bernard Schriever's contributions in transforming American air power to aerospace power.

The famous photograph of Orville Wright at the controls of the *Flyer* on 17 December 1903 forever captures the realization of man's dream to fly. One can only guess what was going through Orville's mind as he pitched down the launching rail and rose into the brisk wind. Although he was only airborne for twelve seconds, that was the beginning of air power and a new, historical chapter in human history. *Winged Crusade: The Quest for American Air and Space Power* celebrates not only this centennial of Orville Wright's flight but also the foundation of an American vision of air and space power. This volume clarifies our understanding of the origins of American aviation and serves to inform its inheritors. In the almost forty-year history of the Air Force Academy symposia proceedings, this is only the third volume dedicated solely to air power. When the Wright brothers succeeded so long ago, they brought a sense of innovation to technology. They connected human spirit to flight and they unknowingly established a new way of thinking about warfare. The essays in this volume trace the development of this new way of thinking—the Winged Crusade—and demonstrate that it must go on.

NOTE

1. <http://www.whitehouse.gov/news/releases/2006/10/20061014-2.html>.

Wright Brothers:
Pattern for American Air and Space Power—
Innovation and Legacy of Ingenuity

Tom D. Crouch

The Twentieth century is over. We prevailed, and we all appreciate the role that air power played in bringing that about. It seems doubly appropriate in a year when America commands the sky as no other nation ever has, that we should also be celebrating the centennial of powered, controlled, heavier-than-air flight, an American achievement that shaped the course of the century. Is that a coincidence? Is there something in our national character, our psyche, that specially suits Americans for aerospace achievement?

It is a tempting notion. Wilbur and Orville Wright were certainly the quintessential middle Americans. Their story seems to exemplify our national strengths and values. They were the boys next door who made good through hard work, commonsense, perseverance, and native ingenuity. In fact, however, it is more useful to see the Wrights as especially talented members of an international aeronautical community. They began their own work on a foundation laid by their European predecessors. Initially, their achievement had a far greater impact in Europe than in America.

Octave Chanute, a Chicago-based civil engineer who was the Wright brothers' closest confidant in aeronautics, had begun to create an informal international community of flight researchers in the early 1880s. Having surveyed the history of the subject, he launched a vast correspondence with aeronautical experimenters scattered around the globe. He identified flight enthusiasts, gathered information from them, offered encouragement, spread news of what was going on elsewhere, and provided occasional financial support. Chanute wrote an authoritative book, *Progress in Flying Machines* (1894), published articles on flight technology in popular magazines and professional journals, lectured widely, and organized sessions of aeronautical papers for leading engineering and scientific societies.

Chanute attracted a new generation of workers into the field. The most notable of these was Samuel Pierpont Langley, who, as Secretary of the Smithsonian Institution after 1887, was effectively the unofficial chief scientist of the United States. In the spring of 1896, Langley and his team launched a steam-powered model aircraft with a fifteen-foot wingspan. The craft remained in the air for 1 minute and 20 seconds, climbing to an altitude of seventy feet and covering a distance of three-quarters of a mile.

In May 1900, Chanute received a letter from yet another newcomer. Wilbur Wright, a thirty-three year old bicycle maker from Dayton, Ohio, admitted to being "afflicted with the belief that flight is possible to man."[1] It was the first of hundreds of letters and telegrams that would pass between the two men over the next decade. Chanute would become the best friend and closest confidant that Wilbur and Orville Wright had in aeronautics. His work would also provide them with their basic approach to an aircraft structure. His most significant contribution to their success, however, was to introduce the Wrights to the technical details of Otto Lilienthal's work.

Lilienthal provided the essential starting point for Wilbur and Orville Wright. They had followed his work in newspapers and magazines for almost a decade before their own entry into the field. Their decision to undertake experiments with manned gliders was certainly inspired by his example. He provided the coefficients for lift and drag, and the algebraic equations that would enable them to use such information to calculate the amount of wing area required to lift a glider and pilot into the air at a given airspeed. Without Lilienthal it is difficult to imagine how or where they would have begun.

All of this is to underscore the fact that aeronautical research in the early twentieth century was an international enterprise. The achievement of Wilbur and Orville Wright was not dependent on the fact that they were quintessentially American. Had they been born with the same talents under similar circumstances in France, England, or Germany, they might still have been the inventors of the airplane. In spite of increasing activity in the United States, the airplane could as easily have been invented in Europe.

Why Wilbur and Orville? How were these two bicycle makers from Dayton able to succeed where so many others had failed? In part, it was a result of their upbringing and character. They had grown up in an extraordinarily tight-knit family, where children were encouraged to explore, experiment, and think for themselves. Their mother, the daughter of a carriage maker, was well educated and good with her hands. Their father, a clergyman whose career was punctuated by a long series of disputes with fellow church leaders, taught his children to have the courage of their convictions and to put their trust and faith in family. Neither Wilbur nor Orville ever married, nor did either of them ever find a better friend than his brother. They knew, understood, and trusted one another. Together as the Wright brothers, the whole was much greater than the sum of the parts.

The Wrights obviously brought special gifts and insights to the process of invention. Had you lived next to them on the West Side of Dayton in the year 1899, you would have regarded them as the most ordinary of men—friendly small businessmen, honest as the day is long, devoted to their family. But you never would have guessed that these two brothers were going to solve the great technical problem of the age, and change the course of history in the process. There were clues, however, for those who knew where to look. The airplane was not their first invention. They had designed and built printing presses and self-oiling wheel hubs for the bicycles that they manufactured. In both cases, they had approached the problems of design from their own unique and unexpected angle, producing a device that sometimes puzzled knowledgeable professionals. "Well, it works," remarked a visiting printer who inspected a press that the Wright brothers had constructed, "but I certainly don't see how it does the work."[2]

Wilbur Wright, in his early thirties, was running two small businesses with this younger brother and living in his father's house. "I entirely agree that the boys of the Wright family are all lacking in determination and push," he admitted to his sister-in-law. "None of us has as yet made particular use of the talent in which he excels other men, that is why our success has been very moderate."[3] Determined to break that mold, he wrote to the Smithsonian on 30 May 1899, announcing his interest in aeronautics and asking for advice on useful readings in the field. Over the next eight weeks, Wilbur and his brother Orville laid a firm foundation for their future, identifying a few kernels of useful information in the work of their predecessors, and carefully analyzing the problem. It was clear to them that an airplane would be a complex machine composed of three systems. It would have to have wings to lift it into the air, a propulsion system to move it forward, and a means of controlling it in the air.

Lilienthal had built wings capable of carrying him through the air on 2,000 flights between 1890 and 1896. Langley's model wings seemed to have worked quite well. The Wrights decided that they were safe in combining their own common sense with Lilienthal's data and his approach to calculating the required wing area. They would be flying gliders, so propulsion would not be an immediate problem. In any case, automobile builders were developing ever lighter and more powerful engines. If and when the time came for a power plant, the technology would be available. That left the problem of control. "When this one feature has been worked out the age of flight will have arrived," Wilbur explained, "for all other difficulties are of minor importance."[4]

Aerodynamic control was the element of the total problem that had received the least attention. Model builders like Samuel Langley had designed inherent stability into their craft, employing wing dihedral and a horizontal tail set at a slightly negative angle to keep the model moving forward in a straight line. As a result, they learned nothing about flight control. Most glider builders had relied on weight shifting control, moving their legs and lower torso in an effort to keep the center of gravity of the machine on top of the center of pressure of the wing. It was a dangerous technique that limited the size of the machine and the extent of control that a pilot could maintain. It had killed both Lilienthal and the English experimenter Percy Pilcher, who died in an 1899 glider crash. The Wright brothers were determined to develop a mechanical system that would enable the pilot, with a few simple movements, to maintain absolute control over an aircraft at all times.

After a false start or two, the brothers came up with the notion of inducing a twist across the entire wing, increasing the angle of attack, and lift, on one side of the machine, and reducing the angle, and the lift, on the other. By manipulating the geometry of the wing in this fashion, they would control the movement of the center of pressure with regard to the center of gravity, maintaining precise control of the entire machine.

In addition, they were decided on a specific structural design—a biplane configuration in which the two wings were linked through a system of struts and wires in a standard engineering truss. In this manner, the relatively frail single wings were transformed into a very strong trussed beam structure. The basic idea was inspired by a well-engineered trussed biplane glider developed by Octave Chanute and his associates in 1896. The Wright brothers added their own brilliant twist to the design, however. Like the Chanute original, they would rigidly brace their biplane along the leading and trailing edges. They would not truss the ends, however. Rather, the wires that would warp, or twist, the wings would be a closed loop, maintaining the strength of the beam across the ends, while at the same time allowing for wing warping. It was brilliant, elegant engineering.

The creation of a system to provide effective lateral control is an example of the extent to which Wilbur and Orville Wright were able to think in non-traditional ways. They had the ability to visualize a machine that had yet to be built, and to imagine how it would behave with forces operating on it. It was only one of a series of gifts that enabled them to succeed where so many others had failed. They had an intuitive grasp of a process that would enable them to move forward toward a full solution to the manifold problems of flight. It was something more sophisticated than the notion of design, test, and incorporate the lessons learned in the next design. Had their approach been that simple, they would not have been able to isolate and study problems in a specific system. Rather, the brothers were able to test specific aspects of their craft. With some notable exceptions, they were careful to incorporate changes in a new design so that the impact of a single alteration could be evaluated.

Wilbur tested the basic wing warping system with a kite built and flown in Dayton in the summer of 1899. Having satisfied themselves that their system of lateral control would work, the next step was to design and build a full-scale machine. Using Lilienthal's table of lift and drag coefficients, they calculated the wing area required to lift the estimated weight of the glider and a pilot in a wind of given velocity. The calculations indicated that the craft would either have to have enormous wings or be flown in a considerable headwind. Correspondence with the U.S. Weather Service regarding average wind speeds across the United States led to the selection of Kitty Hawk, North Carolina, as the site for a "scientific vacation," during which they could test their new craft.

When testing began, it was immediately apparent that the craft was generating far less lift than their calculations had predicted. True to their method, the brothers tied off the flight control system to simplify matters and focus on the central problem at hand. The Wrights devised a means of precisely measuring the forces acting on their machine when being flown as a kite. They attached a grocer's scale to the kite line, which provided a measure of the total force operating on the machine. They measured the angle of attack at which the kite was flying, and determined the wind speed with an anemometer. With that information in hand, and some simple trigonometry, they could calculate the actual lift and "drift," or drag generated by the craft.

Their next machine, the 1901 glider, was both the least satisfactory and most instructive of their aircraft. They made the mistake of introducing too many variables. Unsure as to the source of the aerodynamic problems encountered in 1900, they increased the wing area of their 1901 glider from 165 feet to 290 square feet, adopted a much lower aspect ratio, and introduced a radically different airfoil. While the larger wing area allowed the Wrights to make repeated glides, the aircraft was still delivering 20 percent less lift than the calculations predicted. Because of the multiple changes in wing design, the brothers had no way of isolating the problem, or understanding the impact of any one of the variables.

Worse, now that the Wrights were spending more time in the air, they discovered serious and unexpected control problems. When the pilot warped the wings to increase the lift on one side of the machine, that wing would often slow and drop, rather than rising, sending the aircraft into an incipient spin. The Wrights reasoned that addition of a rudder would balance the increased drag on a positively warped wing and allow effective lateral control. When incorporated into the design of the 1902 glider, the rudder was originally fixed. It was almost immediately made moveable, however, and linked to the wing warping system to increase its effectiveness in countering adverse yaw and allowing for smooth and controlled banks and turns.

At the end of the 1901 season at the Kill Devil Hills, Chanute assured the Wrights that they had moved far beyond all previous experimenters. Indeed, the Wrights were pleased to have spent some time in the air, to have discovered the problem of adverse yaw, and to have come up with a tentative solution that they could test on their next machine. At the same time, they knew that the continuing gap between their calculations and the actual performance of their first two gliders was evidence of a serious underlying problem. While they had not suffered any serious injuries in 1901, the fate of Lilienthal and Pilcher was never far from their minds. "When we left Kitty Hawk at the end of 1901," Wilbur later recalled, "we doubted that we would ever resume our experiments."[5]

Instead, they did a courageous thing, discarding all of the Lilienthal data on which they had based their performance calculations. It was the great turning point in their story. Had they suspected that there were problems with the inherited information at the outset, they might never have begun. They would not have had a firm starting point. Now they set out to gather their own data. They employed a simple device mounted over the handlebars of a bicycle to confirm that the existing data was flawed. The next step was to build a wind tunnel.

Francis Herbert Wenham had built the first such aerodynamic testing device in the 1870s with a grant from the Aeronautical Society of Great Britain. If the Wrights did not invent the wind tunnel, they were certainly the first to produce useful balances, the delicate instruments placed in the air stream inside the tunnel to measure the minute forces playing across the miniature wings being tested. Small enough to fit in a shoebox, constructed of old hacksaw blades and bicycle spoke wire, the two balances—one to measure lift, the other resistance—were brilliantly conceived mechanical analogs of the algebraic equations that the brothers were using to calculate performance. There is no clearer example of Wilbur and Orville's ability to move from the abstract to the concrete, from a mathematical formulation to a brilliantly conceived machine that would provide the precise bit of data required.

Lilienthal had tested a single airfoil shape. In less than a month during the fall of 1901, the Wrights gathered data on the lift and resistance of over forty airfoils through an entire range of angles of attack. In addition, they conducted wind tunnel experiments to determine the best aspect ratio for a wing, the efficiency of various wingtip shapes, and the aerodynamic effect of changing the gap between the wings of a biplane. No longer dependant on borrowed data, the brothers could now move forward with confidence. It was the great turning point.

The 1902 Wright glider was the result of two years of experience in the field, and a few short weeks of path-breaking research with the wind tunnel. Unlike its predecessors, the new glider performed exactly as calculations predicted. The addition of a moveable rudder resolved the control problems of 1901. After three years of effort, the Wrights had achieved their initial goal—to produce a heavier-than-air flying machine operating under the complete control of the pilot. Wilbur and Orville completed perhaps 1,000 glides with the machine between 20 September and 24 October 1902, and continued to fly the glider the following year, while they were assembling and testing their 1903 powered machine.

The Wrights did not patent a powered flying machine. They patented their 1902 glider, reasoning that three axis flight control was a patentable system. No one, they were confident, would be able to fly without drawing on ideas covered by their patent. Their all-important aerodynamic data, which could not be patented, would be kept secret and made available to manufacturers who purchased the right to use their patent.

If Wilbur and Orville had overestimated the reliability of the aerodynamic data gathered by their predecessors, they had also underestimated the difficulties that they would face in the area of propulsion. The engine was not the problem. When a preliminary effort failed to turn up a manufacturer willing to build a power plant to their specifications, they decided to design and build it themselves. Charles Taylor, the machinist whom they employed at the bike shop, produced an engine that did the job. It was not a particularly powerful engine. The key was the fact that the brothers could calculate the power required to fly, and estimate a maximum weight for the engine. Once they had a power plant that met their specifications, they were satisfied. Any attempt to improve the engine beyond what was required, they realized, would represent over engineering, and a waste of time and effort.

The Wrights had not anticipated difficulties with propeller design. Their original assumption that they could borrow a design approach from marine propellers proved incorrect, however. In order to produce propellers whose thrust could be accurately calculated, the Wrights developed a theoretical approach to propeller design that began with a recognition that a propeller was a rotary wing, in which the lift was vectored as thrust. The brothers were familiar with the work of the physicist/engineer W. J. M. Rankine, and used a form of what engineers call blade element theory to design their propellers. Recent wind tunnel tests revealed that first generation Wright propellers are almost 80 percent efficient. That is to say, the propellers transform 80 percent of the engine horsepower delivered to them into thrust. A modern wooden propeller for a light airplane is usually only 85 percent efficient. No other experiment-ers of the era came even close to the efficiency achieved by the Wright propellers. That is what kind of engineers they were.[6]

This year, 2003, we celebrate the centennial of the four powered, sustained and controlled flights that the Wright brothers made from the sand flats at the base of the Kill Devil Hills, four miles south of Kitty Hawk, between 10:35 a.m. and noon on 17 December 1903. Of course that was not the end of the process of invention. On the best of those four flights, Wilbur Wright flew 852 feet over the sand in 59 seconds. The brothers knew that, with practice and an opportunity to keep improving the elements of their machine, they could do much better.

The Wrights decided to continue their experiments at Huffman Prairie, a 100-acre pasture some eight miles east of Dayton By working close to home, they could devote full time to their experiments while keeping an eye on their business and avoiding the expense of living away from home. They worked at Huffman Prairie for the next two years, designing and building a new airplane each year. Initially, they found that the absence of steady headwinds made it difficult to get into the air. Once they began using a catapult system in September 1904, they began to make real progress. By the fall of 1905, they were covering distances of over twenty miles through the air, flying tight circles over their field for over half an hour at a time. They had transformed their marginal success of 1903 into the reality of a practical airplane.

The brothers were never happier than when they were wrestling with a difficult technical challenge. "Isn't it wonderful," Orville wrote to his friend George Spratt, "that all of these secrets have been preserved for so many years just so that we could discover them!!"[7] The process of invention that had begun with a wing-warping kite in the summer of 1899 was now complete, but the career of the Wright brothers was far from over.

The Wrights grounded themselves from 1905 until 1908, while they struggled to market their invention. By May 1908, with signed agreements in hand for the sale of airplanes to both a French syndicate and the U.S. Army, the brothers equipped their 1905 airplane with upright seating for a pilot and passenger and a new upright control system, they returned to the Kill Devil Hills to brush up their flying skills with the new arrangement. Then Wilbur was off to France, where he made his first public flights at the Hunaudières race course, near Le Mans, on 8 August 1908. Orville followed with his first flight to demonstrate their machine to the U.S. Army at Fort Myer, Virginia, on 3 September. For five years, the brothers had been shadowy figures, their claims widely reported in aeronautical circles, and widely doubted. They swept all of the doubts away with these first public flights and emerged as great public heroes on both sides of the Atlantic.

Of course, by 1908, the Wrights no longer had the sky to themselves. Alberto Santos-Dumont, a wealthy Brazilian living in Paris, had flown his airplane, *14-bis,* 722 feet through the air in 22.5 seconds on 12 November 1906. On 26 October 1907, Henri Farman piloted his Voisin aircraft through a full circle, remaining in the air for 74 seconds. It was the first time that anyone had matched the Wright brothers' 1903 performance. In July 1908, just a month before Wilbur made his first flights in France, Farman remained in the air for 20 minutes, 22 seconds at Ghent, Belgium. There was fresh activity in America, as well. On 4 July 1908, Glenn Hammond Curtiss of Hammondsport, New York, won the Scientific American Trophy for the first flight in the United States of more than one kilometer.

It is safe to say that none of those individuals would have flown had it not been for Wilbur and Orville Wright. The brothers were as important to this generation as Lilienthal had been to them. Now they were the starting point. They had met Glenn Curtiss, and answered his technical questions and those of his colleagues in the Aerial Experiment Association, a group organized by Alexander Graham Bell in 1907. In 1900 and 1901, the brothers had published three technical articles, complete with photos of their first two gliders.[8] Chanute had described their work to his wide circle of correspondents, and given a slide-illustrated lecture on the subject to the members of the Aero Club de France in April 1903. The substance of the talk, and the illustrations, were published within the month.

Virtually all of the French aviators who took to the air in 1907–1908 had entered the field as a result of Chanute's lecture/article. Most of the new generation of French aeronautical pioneers began by building their own versions of a Wright glider. Santos-Dumont's *14-bis* and the early machines designed and built by the Voisin brothers and Henri Farman were braced biplanes with pusher propellers and an elevator located in front of the leading edge of the wing. That is not a natural configuration. The French used it because they knew that was what the Wrights had done. None of them understood or appreciated the Wright control system, but they knew what the Wright aircraft looked like, and they knew the Wrights had flown. Make no mistake, the subsequent history of aviation begins with the Wright brothers.

The Europeans who began running to catch up with the Wright brothers swept past them by 1909, and kept right on going. Having mastered flight control after Wilbur's first flights in France, they seized leadership in world aeronautics from the land where the airplane had been born. Louis Blèriot's flight across the English Channel on 25 July 1909, followed by the first great international aviation meet and competition held a month later (22–29 August) on the plain of Bethany, three miles north of the cathedral city of Reims, marked the beginning of European hegemony in the air. The six years remaining before the outbreak of war in August 1914 witnessed constant startling improvements in performance, almost all of them the work of Europeans.

New developments in engine technology were of central importance. By 1914, the four-cylinder, 12.5-HP Wright engine of 1903 had given way to 100-HP eight-cylinder water-cooled in-lines and 90- to 140-HP rotaries like the *Gnôme*. Louis Bechereau had incorporated the monocoque structure, originally developed by the Swiss engineer Eugene Ruchonnet, in the design of the *Deperdussin* racing monoplanes. Hans Reissner experimented with corrugated aluminum wings, while Ponche and Primard produced the *Tubavion* monoplane, the first genuinely all-metal aircraft.

Henri Fabre made the first water take-off on 28 March 1910. The Russian Igor

Sikorsky pioneered very large aircraft with his four-engine *Bolshoi* of 1913. The following year, Glenn Hammond Curtiss produced a multi-engine flying boat intended to fly the Atlantic. The ocean would have to wait for another five years, but, by August 1914, the North American continent had been flown coast to coast, both ways, and both the Alps and the Mediterranean had been traversed by air.

On 17 December 1903, the world's first airplane had traveled a maximum distance of 852 feet in 59 seconds, reaching a speed of 30–35 MPH and an altitude of 10–15 feet. Ten years later, only six years after the Wrights had first flown in public, the records had increased to a speed of 126.67 MPH (Maurice Prevost in a *Deperdussin*); a distance of 634.35 miles over a closed circuit (A. Seguin in a *Farman*); and an altitude of 20,079 feet (G. Legagneux in a *Nieuport*).

The American Glenn Curtiss had won the first James Gordon Bennett race, staged as part of the Reims meet in 1909. By 1913, the United States could not field a competitor for the same race. "We could not send an American biplane or monoplane over," Alan Hawley, President of the Aero Club of America, explained, "because none of our machines are half speedy enough." The airplane, born in America, came of age in Europe.[9]

How did the nation that had given birth to the airplane fall so far behind so quickly? It has often been suggested that the series of patent suits brought by Wilbur and Orville Wright were to blame for the retarded growth of aeronautics in the United States. A careful analysis of the situation is an important first step in understanding the underlying economic and political forces that would drive flight technology for the rest of the century.

The American aviation industry got underway on 3 March 1909, when Glenn Curtiss and Augustus M. Herring incorporated the Curtiss-Herring Company. They quickly sold one airplane to the Aeronautic Society of New York and began entering prize competitions. The Wright Company was incorporated under the laws of the State of New York on 22 November 1909. Wilbur served as president of the firm, and Orville as vice president. The board of directors included August Belmont, Cornelius Vanderbilt III, Robert Collier, and other leading figures in American business and finance. Corporate offices were in New York, but the heart of the operation, the factory and flying school, were in Dayton, where the brothers could retain personal control.

The new factory began production in November 1910, turning out Wright Model B aircraft. Powered by a 40-HP engine, it was the first Wright production aircraft, and the first Wright machine mounted on wheels. When operating at full capacity, the workmen could push two airplanes a month out the factory door. The company produced twelve distinct aircraft designs between 1910 and 1915, when Orville Wright sold his interest in the firm. While precise figures are not available, the Model B and Model C, which sold to the U.S. Army, were produced in the largest numbers. Other models included the EX, which Calbreath Perry Rodgers flew from coast to coast in 1911; the Model R, designed for air racing; and the Model G flying boat.

Even before the founding of the company, the Wrights and their backers had recognized that the basic Wright patent, granted in 1906, was one of their most valuable assets. The era of the patent suits began 18–19 August 1909, when the Wrights filed a bill of complaint enjoining their principal American rival, Glenn H. Curtiss and the Herring-Curtiss Company, from the manufacture, sale, or exhibition of airplanes that infringed on the Wright patent.

The patent litigation spread to Europe in 1910, when the Wright licensees, the Compagnie Generale de Navigation Aerienne (CGNA), brought suit against six rival aircraft manufacturers (Blèriot, Farman, Esnault-Pelterie, Clement-Bayard, Antoinette, and Santos-Dumont) for infringing on the Wright's French patents. The following year, a consortium of five German aircraft builders brought suit against the incorporators of the German Wright Company in an effort to overturn the Wright patents in that nation.

The patent suits proved to be difficult, expensive, and time-consuming for all parties. In the end, they failed to produce a clear-cut resolution. In Germany, the courts invalidated the Wright patents, arguing that prior disclosure, the publication of information on the basic elements of the Wright airplane before the approval of their patent, had compromised their claims. The French suit, complicated by a very different legal system and the absence of spirited prosecution by the CGNA, was still not fully resolved when the Wright's French patents expired in 1917.

The situation in the United States was just as complex. As early as 3 January 1910, Judge John R. Hazel of the U.S. Circuit Court at Buffalo, New York, had issued an injunction prohibiting Glenn Curtiss from the manufacture or sale of aircraft. Curtiss posted a $10,000 bond and appealed the decision. He could legally continue flying until the appellate court reached a decision, but he took a terrible risk in doing so. If Judge Hazel's decision was upheld, Curtiss would have to negotiate a settlement with the Wrights covering all of the monies earned while the injunction was in effect. Curtiss moved forward with the prospect of financial ruin staring him in the face.

On 13 January 1914, the judge of the U.S. Circuit Court of Appeals of New York ruled in favor of the Wrights. Rather than taking immediate financial vengeance against Curtiss, the leaders of the Wright Company, sensing the opportunity for monopolistic profits, announced the schedule of rates that they would charge anyone who wished to exhibit an airplane in the United States. Glenn Curtiss, represented by the best lawyers that money could buy, announced that he would immediately alter the control systems of his aircraft so that they no longer infringed on the Wright patent. Few knowledgeable individuals believed that was possible, but it was enough to muddy the waters and set the legal process in motion once again.

Ultimately, Orville did profit from the patent suits. He sold his interest in the company to a group of New York financiers in 1915 for an undisclosed sum said to have been in the neighborhood of $1.5 million. Certainly, it appeared that the patents might enable the firm to dominate the new industry. Orville sold out at the perfect moment, acquiring a personal fortune that would enable him to live comfortably for the rest of his life. Two years later, in 1917, industry leaders, with the support and advice of the federal government, brought the era of the patent trials to an end by purchasing the rights to all aeronautical patents and creating a pool of leading manufacturers who would share access to all patents.

Did the long battle over patents retard the growth of American aeronautics? Evidence to the contrary begins with the fact that the Wright Company was much more severely damaged by the patent suits than the Curtiss Aeroplane and Motor Company. Consider the matter of aircraft sales, the most basic measure of corporate success. Between 1909 and 1915, Wilbur and Orville Wright and the various Wright companies operating on the basis of their patents sold a total of 14 aircraft to the U.S. Army, their largest single customer.[10] Orville Wright estimated the total production of the Dayton factory between 1910 and 1915 at roughly 100 airplanes.

During the same period, the companies controlled by Glenn H. Curtiss sold a grand total of 232 aircraft to the U.S. Army. This number, representing twenty-four distinct designs, was almost half of the total number of aircraft purchased by the Army prior to U.S. entry into World War I, and nearly ten times the number of Wright aircraft purchased during this period.[11] In addition, 16 of the first 27 aircraft purchased by the U.S. Navy were Curtiss machines. The Burgess Curtiss Company, unrelated to Glenn Curtiss, produced 4 of those original naval aircraft. The Wright brothers were in third place with the sale of only 3 machines to the Navy. After 1913, Curtiss sales to the Navy skyrocketed, while the Wright Company sales to the Navy were at an end. The precise figures for civil and foreign aircraft sales are not available, but Curtiss's success in marketing single- and multi-engine flying boats to several Allied nations suggests that he was more successful in those categories as well.[12]

Curtiss prospered during the patent suit era, while the Wrights suffered. At the end of the period, Glenn Curtiss was the most successful producer of aircraft in the United States. He was the principal supplier of training aircraft to the U.S. government, and the only American manufacturer producing combat aircraft of his own design for the Allies. The Wright brothers were out of the airplane business.

Why did the Wright Company suffer as a result of the patent suits? The reasons are not so difficult to understand. Wilbur and Orville Wright, the engineering geniuses at the heart of the firm, paid far more attention to winning victory in the patent suit than to the development of new and improved products. In truth, the brothers wanted most of all to be recognized as the true inventors of the airplane, and for the world to appreciate the magnitude of their accomplishment. Victory in the patent suits, and any money that resulted, would symbolize the realization of those goals. Glenn Curtiss, on the other hand, wanted nothing more than to develop, build, and sell improved aircraft. He had bet that good lawyers would see him safely through the patent suits. While the patent suits may have frightened some embryonic U.S. aircraft firms, the fact that Curtiss won his bet suggests that the patent suits had at best a limited impact on the development of aviation in the United States.

If the patent suits do not explain the retarded growth of the industry in America, what forces were at work? Before World War I, the pressure of competition was an important factor encouraging technical progress. Initially, there was little to differentiate the prizes and rewards available to aviators in Europe and America. That began to change by the summer of 1909, as the leading European powers sought to showcase the aeronautical achievements of their citizens. Having served as the site of repeated competitions, cities like Blackpool, Hendon, Reims, Milan, Vienna, and Berlin emerged as world aviation centers. Between May 1910 and October 1913, Johanisthal, the principle Berlin flying field, hosted a total of seven *Flugwoche* (flying weeks), offering a total of 312,900 marks in prize money. In addition, the field served as either the starting point or an important stop on a number of famous long-distance contests, including the Circuit of Germany (12 June–10 July 1910); the Berlin to Vienna Race (9 June 1912); and the Circuit of Berlin (31 August–1 September 1912).[13]

The more strenuous competition and richer prizes available in Europe fueled technological change. In the United States, the leading aviators were members of two or three touring exhibition teams who earned salaries for performing aerial stunts to thrill crowds of paying customers. There were no better pilots in the world than men like Lincoln Beachey and Walter Brookins, but they did not face the constant pressure to fly higher, faster, and farther against a wide range of competitors, week after week. More important, their technology was not tested either. The original configuration of the Wright airplane—a pusher biplane with a canard elevator—remained the U.S. standard until 1911, when a series of accidents among military aviators led companies like Curtiss and Martin to switch to the tractor configuration.

The threat of war was an even more important factor. No one was certain what military utility the airplane might have. Faced with rising international tensions during 1900–14, however, European leaders could ill-afford to allow a rival nation to forge ahead with the new technology. "With Russia and Austria-Hungary in their present troubled condition, and the German Emperor in a truculent mood," Wilbur Wright noted, "no government dare take the risk of waiting to develop practical flying machines."[14]

As a result, European governments encouraged the development of a domestic aircraft industry through investment and subsidies. The scale of European investment in flight technology is apparent in Table 1 listing national spending for aeronautics prepared by U.S. Army officials in 1913. In addition to official appropriations, several leading aeronautical powers had also established national subscriptions that provided an additional $7,100,000 in financial support for their aeronautical industries. Once again, Germany led the way with $3,500,000 in private funds, followed by France ($2,500,000), Italy ($1,000,000), and Russia ($100,000). According to official U.S. government estimates, the nations of the world had spent a total of $93,620,000 in public and private funds on aviation during 1908–13.[15]

How was this money spent? England, France, Germany, and Russia invested large sums on state-supported, well-equipped aeronautical research facilities. Patriotic philanthropists also contributed to the cause. In France, for example, industrialist Henri Deutsch de la Meurthe, Gustave Eiffel, and Basil Zaharoff pursued aerodynamic research, created aerodynamic institutes at French universities, and established endowed chairs in aeronautical engineering.[16]

In addition to government support for flight research, European nations nurtured aircraft manufacturers. The impact of this funding was dramatic. By 1914, the

Table 1. Total Government Expenditures on Aviation, 1908-1913

Country	Expenditure in U.S. Dollars, 1913
Germany	$28,000,000
France	$22,000,000
Russia	$12,000,000
Italy	$8,000,000
Austria	$5,000,000
England	$3,000,000
Belgium	$2,000,000
Japan	$1,500,000
Chili	$700,000
Bulgaria	$600,000
Greece	$660,000
Spain	$550,000
Brazil	$500,000
United States	$435,000
Denmark	$300,000
Sweden	$250,000
China	$225,000
Romania	$200,000
Holland	$150,000
Serbia	$125,000
Norway	$100,000
Turkey	$90,000
Mexico	$80,000
Argentina	$75,000
Cuba	$50,000
Montenegro	$40,000

Source: *Aeronautics in the Army,* Hearing before the Committee on Military Affairs, House of Representatives, Sixty-Third Congress, First Session (Washington, D.C.: Government Printing Office, 1913)

Farman company employed 1,000 individuals in a series of plants scattered around Paris.[17] During the years 1909 to 1914, Blèriot Aeronautique produced 800 airplanes. The fact that two major firms, Deperdussin and Nieuport, prospered in spite of the early loss of very strong founders, is striking evidence of a growing industrial maturity. European governments at least occasionally targeted spending to support a troubled company. In April 1910, for example, when a flood devastated the Voisin factory, the French government ordered thirty-five aircraft from the firm in a successful effort to prevent a collapse.[18]

The European propulsion industry also prospered during the years leading up to World War I. In 1913 alone, the 650–800 individuals employed at the Gnome factory at Gennevilliers produced a total of 1,400 rotary engines. Renault, the second largest French producer of aircraft engines, provided fully one-third of the power plants purchased by the French military. The French aeronautical engine industry, the world leader by 1914, combined the use of the latest American machine tool technology with the older French tradition of hand-crafted excellence in the metal trades.[19]

Without the incentive of war looming on the horizon, the United States did not even attempt to keep up. In 1910–11, a period during which the U.S. Army took delivery of 14 airplanes, the French government ordered over 200 flying machines. In 1912, a committee of wealthy French patriots raised four million francs to supplement the national budget for military aviation.[20] That year the U.S. Secretary of the Navy pointed out that France had spent $7,400,000 on flight to date. Russia had spent $5,000,000; Germany $2,250,000, and Great Britain and Italy $2,100,000 each. Even Japan ($600,000) had out spent the United States ($140,000).[21] By 1913, fourteen nations—including Belgium, Japan, Chili, Bulgaria, Greece, Spain, and Brazil—were spending more on aviation than the United States.

European nations recaptured the lead in flight technology, not because the Wright patent suits had retarded American development, but because European government investment had fueled the rapid development of aviation. The gap grew even wider as the pressure of war further accelerated developments in Europe. The result was, of course, that American airmen flew off to war in 1917 in aircraft that had been almost entirely designed and built, in Europe.

By that time, the Wright brothers were no longer an active force in aeronautics. Wilbur had died of typhoid in 1912. The members of the family blamed it on the stresses and strains of the patent suits. Having sold the Wright Company in 1915, Orville allowed the Dayton-Wright Company to use his name, and served as a consulting engineer with the firm during World War I. He made his last flight as pilot in command on 13 May 1918. He would remain an honored figure, showered with honorary degrees and other awards and honored at scores of dinners and ceremonies for the rest of his life. The Wright achievement was commemorated by a great national monument dedicated in 1932, overlooking the spot where the brothers had made their first successful sustained and controlled powered flights. President Woodrow Wilson appointed him to membership on the National Advisory Committee for Aeronautics in January 1918, a position he held until his death on 30 January 1948.

By then, the United States was the world's dominant air power. The American aircraft industry had finally achieved parity with Europe during the years immediately following World War I. There things remained until the design revolution of the mid-1930s. The new generation of streamlined, all-metal aircraft that took to the air in those years were not the result of any one breakthrough. Rather, they were the end product of a wide range of innovations in structures, aerodynamics, and propulsion that had occurred over a period of at least fifteen years.

The process of change began with the development of the first metal aircraft prior to World War I, continued with the introduction of duralumin, and culminated with the appearance of anti-corrosive coatings and new machine tools and production techniques. Variable speed and constant-pitch propellers enabled the new aircraft to make full use of powerful radial engines shrouded in drag-reducing cowlings. Stressed skin cantilevered wings, streamlining, the retractable landing gear, and flush riveting reduced drag, while high-lift devices enabled the new all-metal, high-performance monoplanes with their high-wing loading to take-off and land safely. Each innovation had been significant in its own right. When engineers combined them to create a new generation of aircraft during the years 1933–35, they added up to a revolution in aircraft design and performance. It was an international revolution. Streamlining, monocoque construction, and other structural innovations appeared in France before 1914. Metallurgical developments, experience in metal aircraft design, and theoretical research in areas from aerodynamics to structures had flowed from Germany to other parts of Europe and America. English engineers took the early lead in radial engine design, and provided such key innovations as the sodium-cooled valve and the first drag-reducing cowling. An English aircraft won the Guggenheim Fund Safe Airplane Prize by demonstrating the use of the slats and flaps that would enable larger, heavier aircraft to take-off and land safely.

While the impact of the design revolution was apparent in new aircraft produced in many nations, it was obvious that American industry was most successful at integrating new technologies into successful aircraft. New American airliners like the Douglas DC-3 set a world standard. The enormous industrial expansion of the U.S. aircraft industry during World War II completed the process begun in the 1930s.

America did not have a monopoly on good ideas or brilliant engineers. The European origins of the jet engine and the large ballistic missile were proof of that. The strength of the American economy, heavy postwar government spending on all aspects of the aerospace enterprise, a strong commercial airline system, a thriving military and civil space program, and other factors insured that those ideas would find their fullest expression in the United States. The success of the U.S. aerospace industry has been a major factor in our success as a nation since 1945. Through the end of the 1970s, U.S. manufacturers supplied perhaps 75 percent of the world's large commercial airliners. By the end of the century, as noted at the outset, the success of American air power enabled the nation to control the airspace over any potential battlefield.

As we look forward into the twenty-first century, problems cloud the horizon. The very fact of American power inspires bitter opposition. Global competition challenges traditional American mastery of the market for large commercial aircraft. Analysts point out that our high-tech workforce is aging. And among the many things that we learned on 11 September 2001 was the relative ease with which adversaries who are weak in traditional military terms can turn the most sophisticated products of flight technology against us. The terrorist hijackers paid absolutely no attention to the good intentions of the men and women who design, build, and operate modern airliners. The words of historian Melvin Kranzberg ring truer than ever. "Technology is neither good, nor bad," he explained, "nor is it neutral."[22]

For better or worse, flight technology is ours to do with what we will. A thousand years from now, when our descendants look back on the twentieth century, they will surely remember this as the time when human beings first took to the sky. Whatever its near-term consequences, flight has had a profound psychological impact on us.

The Wright brothers launched a new era in history—the age of flight. I work in a museum that is a shrine to the air age. In an average year, 9 million people will walk through the doors of the National Air and Space Museum (NASM)—14 million in our best year. We welcome more visitors than the British Museum, the American Museum of Natural History, the Metropolitan, or the Louvre. It is the most visited museum in the world. When NASM opened to the public on 1 July 1976, the staff was confident of success, but no one expected the enormous number of visitors who arrived that first summer, or the wave of media enthusiasm that washed over the building. President Gerald Ford commented that the museum was "our bicentennial birthday present to ourselves." In fact, those of us who planned the museum could take only limited credit for its success.

The quality of the NASM collection is a far more important factor. What other museum in the world, covering any subject, can offer such riches? Visitors to the NASM can see the world's first airplane; the world's first military airplane; the first airplane to fly around the world; the *Spirit of St. Louis;* the Lockheed Vega that Amelia Earhart flew across the Atlantic; Wiley Post's *Winnie Mae*; Howard Hughes's classic H-1 racing aircraft; the B-29 *Enola Gay*; the Bell X-1 that Capt. Charles Yeager, he of the right stuff, first flew faster than sound; the world's fastest airplane; the first airplane to fly around the world non-stop and unrefueled; the first balloon to circumnavigate the globe; the first helicopter to fly around the world; the world's oldest liquid propellant rocket; the spacecraft that carried the first American into orbit; and the Apollo 11 Command Module that brought the first human beings to walk on the surface of another world home again. And that is only the tip of our iceberg.

But the core of the museum's appeal runs deeper even than the opportunity to see the actual aircraft and spacecraft in which intrepid men and women wrote the history of the twentieth century in the sky. However one assesses the immediate consequences of aviation, flight remains one of the most stunning and magnificent of human achievements. People flock to the NASM from around the world because this is a museum that makes them feel proud to be human.

That is the legacy of Wilbur and Orville Wright. They were the inventors of the airplane in a much truer sense than Thomas Edison can be said to have invented the light bulb or Alexander Graham Bell the telephone. The Wrights chose the most difficult technical problem in sight. They analyzed the complex and confusing record of pervious experiments in the field, focusing their attention on flight control. Having selected a starting point, the brothers demonstrated their innate sense of a process of invention. They had to learn to fly gliders, risking life and limb, as part of that process. When their gliding experiments failed to provide necessary information, they developed a wind tunnel balance, an elegant engineering instrument that opened the way to success. In six short years, Wilbur and Orville Wright had produced an invention that would define the course of the twentieth century.

The achievement of heavier-than-air flight involved nothing more nor less than the realization of the oldest and most potent of human symbols. To fly is to escape restraint, soar over obstacles, and achieve mastery and control of our fate. Suddenly, the old dream, which had become the very definition of the impossible, was a reality. If human beings could fly, what could they not accomplish? The invention of the airplane threw open the doors to unimagined possibility. That, it seems to me, is an achievement worth celebrating.

Notes

1. Wilbur Wright to Octave Chanute, 13 May 1900, Marvin W. McFarland, ed., *The Papers of Wilbur and Orville Wright* (New York: McGraw-Hill, 1953), 15.

2. Charles J. Bauer, "Ed Sines: Pal of the Wrights," *Popular Aviation* (June 1938), 40.

3. Wilbur Wright to Lulu Billheimer Wright, 18 June 1901, box 7, Papers of Wilbur and Orville Wright, Manuscript Division, Library of Congress.

4. Wilbur Wright, "Some Aeronautical Experiments," *Journal of the Western Society of Engineers* (December 1901).

5. Wilbur Wright, "Brief and Digest of the Evidence for the Complainant. . . ." The Wright Co. vs. the Herring Curtiss Company and Glenn H. Curtiss in Equity, National Archive.

6. Robert L. Ash, Stanley J. Miley, Drew Landman, and Kenneth Hyde, "Evolution of Wright Propellers Between 1903 and 1912," American Institute of Aeronautics and Astronautics Paper #2001-0309; see also propeller information on the Wright Experience website: <http://www.wrightexperience.com/pdfs/props.pdf>.

7. Orville Wright to George Spratt, 7 June 1903, McFarland, ed., 313.

8. For all of the Wright's published articles, see Peter Jakab and Rick Young, *The Published Writings of Wilbur and Orville Wright* (Washington, D.C.: Smithsonian Institution, 2000).

9. Henry Serrano Villard, *Contact!: The Story of the Early Birds, Man's First Decade of Flight From Kitty Hawk to World War I* (New York: Bonanza, 1968), 192.

10. James C. Fahey, *U.S. Army Aircraft (Heavier-Than-Air), 1908–1946* (New York: Ships & Aircraft, 1946), 6.

11. Ibid.

12. Precise statistics on early U.S. military aircraft purchases are difficult to find. The best sources include: Fahey, *U.S. Army Aircraft;* Robert B. Casari, *U.S. Army Serial Numbers and Orders, 1908–1922 Reconstructed* (Chillicothe, Ohio: Military Aircraft Publications, 1995); *United States Naval Aviation, 1910–1980* (Washington, D.C.: Naval Air Systems Command, NAVAIR 00-SOP-1), appendices 4 and 8; William T. Larkins, *U.S. Naval Aircraft, 1921–1941* (New York: Orion, 1988), 344–48.

13. For details on European air meets and competitions, see Villard, 87–160.

14. Fred C. Kelly, ed., *Miracle at Kitty Hawk: The Letters of Wilbur and Orville Wright* (New York: Farrar, Straus & Young, 1951), 170.

15. *Aeronautics in the Army,* Hearing before the Committee on Military Affairs, House of Representatives, Sixty-Third Congress, First Session (Washington, D.C.: Government Printing Office, 1913).

16. Walter T. Bonney, *The Heritage of Kitty Hawk* (New York: Norton, 1962), 155.

17. *L'Aerophile,* no. 6 (15 Mar. 1914), 124–27.

18. Ibid., 160.

19. John Morrow, *The Great War in the Air: Military Aviation from 1909–1921* (Washington, D.C.: Smithsonian Institution, 1993), 33.

20. John H. Morrow, Jr., *The Great War in the Air: Military Aviation from 1909 to 1921* (Washington, D.C.: Smithsonian Institution, 1993), 13. Professor Morrow provides the definitive account of the growth of the airframe and engine industry in the United States and Europe prior to and during World War I.

21. Bonney, 156.

22. Melvin Kranzberg, "Technology and History: 'Kranzberg's Laws.'" *Technology and Culture* 27 (1986):544–60.

Who's in Control?
A Century of Organizing for Air War

Herman S. Wolk

The question of how to organize, control, and employ air power is as old as the air weapon itself. It is not solely a technological issue, being also bound up in politics and service roles and missions. The advance of twentieth-century aviation technology forced dramatic change in the way wars were fought. Consequently, the question of how to organize air forces to implement national objectives took on great urgency and significance.

National air power is dependent upon how air forces are organized and employed. What is air power? According to the *United States Air Force Dictionary,* it is "this power adapted to military ends, the power which a nation may use in war against another nation or nations, based upon a national weapons system organized about air forces."[1] General Henry H. (Hap) Arnold, architect of twentieth-century air power, defined it as "a nation's ability to deliver cargo, people, destructive missiles, and war-making potential through the air to a desired destination to accomplish a desired purpose."[2]

The problem of air organization has been a constant for almost a century, even after September 1947 when the United States Air Force was established by the National Security Act of 1947. Within the United States Army during the interwar years, airmen chafed over an organization which they thought short-changed the Army air arm. Arguing that airmen knew best how to structure and operate air forces, they fought for their own promotion list, a separate budget, and the opportunity to state requirements directly to national authority without going through the War Department General Staff.

Air forces in World War I demonstrated basic air missions—observation, support of ground forces, even strategic bombing. Major Carl Andrew Spaatz emphasized in November 1923 that "the attributes and advantages of air power can be exploited to the fullest extent only by men fully cognizant of those attributes and advantages, that is to say, by airmen. The best interests of the nation demand of air power an equal place and voice in the councils of war."[3]

This issue was perhaps best stated in 1937 by Major General Frank Maxwell Andrews, a founder of the Air Force, then commanding the General Headquarters Air Force: "I don't believe any balanced plan to provide the nation with an adequate, effective Air Force . . . can be obtained, within the limitations of the War Department budget, and without providing an organization individual to the needs of such an Air Force. Legislation to establish such an organization will continue to appear until this turbulent and vital problem is satisfactorily solved."[4]

The War Department and the Army airmen construed different lessons from World War I. To Army Chief of Staff General Peyton March, the war was won on the ground, not by "some new, terrible development of modern science."[5] Despite the Air Corps Act of 1926, the War Department continued to believe that the primary role of the Air Corps was support of the ground army. However, the Air Corps Act provided for air representation on the General Staff although the airmen remained subject to War Department control. As General Ira C. Eaker retrospectively described it: "We were just sort of voices in the wilderness; a great many military people considered us crackpots."[6] Brigadier General William Mitchell's call for formation of a Department of Aeronautics and his subsequent court martial made news, but obscured much of his thinking. He was ahead of his time, a conceptual air strategist, and unusually competent in planning for organization and control of air forces.

The period between the wars was distinguished by advancing aviation technology and many attempts by congressional advocates to legislate autonomy for the Army's air arm. However, the airmen had no opportunity to demonstrate their theories in combat. Despite a series of legislative setbacks in the 1920s and 1930s, when congressional boards posited that ground support remained the air arm's major function and thus saw no reason to support independence, the Baker Board report of 1934—although opposed to a separate air arm—led in March 1935 to formation of the General Headquarters (GHQ) Air Force commanded by Brigadier General Frank M. Andrews.

James H. Doolittle, a major in the Air Corps reserve, in a dissent to the Baker Board, emphasized: "I believe that the future security of our nation is dependent upon an adequate air force. This is true at the present time and will become increasingly important as the science of aviation advances. I am convinced that the required air force can be more rapidly organized, equipped, and trained if it is developed as an entirely separate arm. If complete separation is not the desire of the committee, I recommend an air force as part of the Army but with a separate budget, a separate promotion list, and removed from the control of the General Staff."[7]

Prior to creation of GHQ Air Force, air units had been under the control of the nine Army Corps areas throughout the country. Formation of this air striking force constituted an important step in evolution of American air power as it centralized operational command over all Air Corps combat units under a single airman.

Nonetheless, the questions of the proper functions and control of air forces persisted. The GHQ Air Force reported to the Army's Chief of Staff and remained a tenant on bases. The Air Corps was split between the office of the Chief of Air Corps (OCAC) and GHQ Air Force. The Air Corps 1936 Browning Board report concluded that "the present organization is unsound" and recommended that GHQ Air Force be consolidated under the OCAC.[8] Subsequently, the War Department exempted Air Corps stations from corps area control.

The fact remained, however, that in the interwar period the airmen could not demonstrate the efficacy of their theories in combat. And the War Department General Staff did not have the ability to bridge the theoretical gap and conceptually vault into the future. Thus it failed to foresee the impact of military aviation upon future conflict. "The treatment of the Army Air Corps prior to World War II by Army decision makers," observed General Jacob E. Smart, one of the finest minds in the Air Force, "stemmed from their perceptions of how the next war would be fought and their limited understanding of the potential capabilities of air power. Those conscientious men were the products of their respective experiences, education, and imagination. They were unable to foresee air warfare becoming significant other than as a supporter of ground warfare and were skeptical of the airmen's assertions about potential air capabilities."[9]

The problem of air organization turned critical during 1939–41. As General Arnold noted, in the 1930s, "air power was the unseen guest at those grim conferences which marked the Nazi rise to power."[10] It was a time when—with the Wehrmacht in full throttle in Europe—President Franklin D. Roosevelt, deeply concerned about the pivotal role played by the *Luftwaffe* in the European war, directed an enormous air buildup while General Arnold desperately attempted to shift the Army Air Forces (AAF) "from low to second gear."

Roosevelt had directed an immense, if unrealistic, aircraft production program. Assistant Secretary of War for Air Robert A. Lovett protested to General Arnold: "It is a little bit like asking a hen to lay an Ostrich egg. It is unlikely that you will get the egg, and the hen will never look the same."[11] Arnold replied, "that if we can induce her to lay it, I, for one, feel that we must accept the wear and tear on the hen."[12]

In early 1941, American-British conversations (ABC-1), crafting objectives in the event of U.S. entry into the war, led to the formation in August of the Combined Chiefs of Staff (CCS), British and American, with General Arnold representing Army air power. Arnold's presence in the CCS—and the Joint Chiefs of Staff (JCS)—reflected the Army Air Forces virtual autonomous position. Subsequently, in 1943, with creation of the joint committee system within the JCS, the Army's official history noted that the AAF "exercised great influence" within committees as to how Army air units were employed.[13]

On the eve of war, the problem faced by Marshall and Arnold was twofold: First, to streamline the General Staff to more quickly build up air power; and secondly, to reorganize to foster efficient and effective wartime operations, should the United States become involved. Among objectives of this reorganization would be providing Arnold and the Air Staff (established in June 1941) sufficient clout and flexibility to be able to move their requirements through the War Department General Staff. As recognized by Marshall, the General Staff had "lost track of the purpose of its existence. It had become a huge, bureaucratic, redtape ridden, operating agency. It slowed down everything."[14] In Marshall's view, the War Department had become "the poorest command post in the Army." Arnold and Robert A. Lovett, both of whom worked well with Marshall, pressed for Arnold to handle all air matters. Marshall, who thought the General Staff was openly hostile to airmen, had earlier named Arnold as acting Deputy Chief of Staff for Air and Major General George Brett as acting Chief of the Air Corps.

Congress meanwhile, turned up the heat, indicating that it might once again attempt to legislate air independence. At this point, Secretary of War Henry L. Stimson stepped in, believing that the Army's air element required great flexibility and freedom to build up its forces to meet the threat of war. Stimson agreed with Lovett and Arnold that air warfare involved independent operations divorced from land and sea action. "The difficulty," Stimson emphasized, "is finding just how far to go" in giving the airmen more freedom. This confluence of thought proved decisive because Stimson was under heavy pressure from Congress to grant more freedom to the Air Corps. In May 1941, Marshall informed Stimson that a revision of Army Regulation 95-5 was ready for implementation. "It thus gave me something with which to meet the threat of an independent Air Corps created by (Congressional) legislation," said Stimson.

Another proponent was Andrews, who in early 1941 was Commanding General, Panama Canal Air Force. He maintained that the Army's air arm could not be properly developed "under an organization which considers it an adjunct of surface forces, even with a man as broadminded and farseeing as Marshall at the head of the Army." He added, "No matter how progressive Marshall may be himself, the rank and file of the Army had not changed materially." Andrews, who did not always see eye to eye with Arnold, nonetheless considered him a good politician and was confident that Arnold could handle this issue.

Having gotten a green light from Stimson, Marshall on 20 June 1941 put into effect revised Army Regulation 95-5, redefining the organization and functions of the Air Corps and officially establishing the Army Air Forces. It gave Arnold the title of Chief, AAF (he continued to be deputy chief of staff for air), responsible to the Army Chief of Staff and the Secretary of War. Under 95-5, Arnold had the authority to coordinate the Office of the Chief of the Air Corps (Major General George Brett) and Air Force Combat Command (Lieutenant General Delos C. Emmons), redesignated from the GHQ Air Force and which previously had reported directly to Marshall. Combat Command would develop air doctrine and plans for operational training. The Chief of the Air Corps would supervise research and development, supply and maintenance. The revised regulation provided Arnold with an Air Staff to formulate policy and plans.

Arnold saw the creation of the Army Air Forces as another step toward independence. The newly formed Air Staff still answered to the War Department General Staff. And the airmen did not have their own budget and promotion system. Relations between the Air Force Combat Command and the AAF continued to be unsatisfactory just as those between the Chief of Air Corps and GHQ Air Force had been divisive. Thus, the expansion of the AAF in response to the demands of wartime led Arnold and Marshall to postpone serious consideration of air independence until after the war. The enormous pressure of wartime requirements, with the attendant breakdown of the General Staff, led Marshall and Arnold to reorganize to allow airmen sufficient flexibility to organize and control air forces for global warfare.

The so-called Marshall Reorganization of March 1942, called "the most drastic and fundamental change" experienced by the War Department since creation of the General Staff in 1903, gave the AAF *de facto* autonomy. Moreover, it allowed the

Army airmen to better control air forces in the theaters of war. Just prior to the Japanese attack on Pearl Harbor, General Arnold had emphasized a new role for air power: "The development of the Air Force as a new and coordinated member of the combat team has introduced new methods of waging war. Although the basic Principles of War remain unchanged, the introduction of these new methods has altered the application of those Principles of War to modern combat."[15] In the past, the military commander has been concerned with the employment of a single decisive arm, which was supported by auxiliary arms and services. Today the military commander has two striking arms. These two arms are capable of operating together at a single time and place, on the battlefield. But they are also capable of operating singly at places remote from each other. The great range of the air arm makes it possible to strike far from the battlefield, and attack the sources of enemy military power. The mobility of the air force makes it possible to swing the mass of that striking power from distant objectives to any selected portion of the battlefront in a matter of hours, even though the bases of the air force may be widely separated.

As noted, the significance of air power's role was also reflected in Arnold's position on the Joint Chiefs of Staff and the Combined Chiefs of Staff. During the war, General Arnold pointed to this development: "The AAF are being directly controlled by the Joint Chiefs of Staff and the Combined Chiefs of Staff more and more each day. Consequently, AAF representation in the joint and combined planning staffs has become a position of paramount importance to me."[16] Control of long-range air operations was of utmost importance to Arnold. As the official Army History noted: "The success of the makeshift organizational arrangements in World War II did not conceal the ultimate importance in future national defense at arriving at a clear—cut definition of the functions and status of the Air Forces in relation to both the Navy and the rest of the Army."[17]

With formation of the Twentieth Air Force in April 1944, Arnold succeeded in establishing independent command in the Pacific—free from control of the theater commander—when the JCS designated the Commanding General, AAF as "executive agent" to direct the long-range B-29 campaign against Japan. Thus, in the Pacific, the AAF gained equality with the ground and sea forces. And in January 1944, the U.S. Strategic Air Forces in Europe had been formed under the CCS to direct long-range operations of the Eighth and Fifteenth Air Forces. Operational control under the Combined Chiefs was vested jointly in the Chief of Air Staff, the Royal Air Force, and the Commanding General, AAF.

Thus, World War II demonstrated that unified command of land, sea, and air forces was absolutely necessary to promote military efficiency. By unified command we refer to component commanders in the theaters of war reporting to a theater commander with a joint staff, responsible to the Joint Chiefs. The war also proved beyond doubt the maturity and importance of air power. Prior to the war, air power critics had pointed out that the claims of air advocates amounted to nothing but theory. That argument died in the wartime skies over Europe and the Pacific. Commanders controlled joint forces and the results were especially impressive in the

European theater. In the Pacific, unified command was not established and a contentiousness existed between the Army and Navy throughout the war.

Interestingly, in the midst of the war, General Douglas MacArthur, Southwest Pacific theater commander, while praising General George C. Kenney, his air commander, noted that evolution and control of the air weapon was "still in its infancy and fifty years from now the world will look back on today with something akin to amazement at its air antiquity. The effect on war and indeed the impact on civilization as a whole of this new element has yet to be comprehended. Its destructive qualities are becoming understood and applied but its logistic possibilities are as yet only mildly realized or utilized."[18]

With great foresight, starting immediately after Pearl Harbor, Arnold formed groups within AAF Headquarters to plan for postwar air organization. Also, in the midst of the global conflict, Congress continued to study the question of military organization, especially as it related to the air element. In the immediate postwar period, airmen echoed themes that had been sounded since post–World War I and the interwar years; that efficient and effective air operations and air-ground integration required an Air Force led by airmen—with appropriate authority and support—who understood how to organize and operate air forces. Direct command of air forces must be exercised by the air commander. In the years immediately after the war, these propositions fell on sympathetic ears. The war also emphasized the importance of long-range air operations which raised unique organizational and doctrinal challenges.

Despite Navy reluctance, President Harry S. Truman, the War Department, and the Army Air Forces were determined not only to create a United States Air Force, but also to build an integrated national security structure. Truman favored "air parity," emphasizing the AAF's contribution to victory in World War II. "In operations," Truman stressed, "air power receives its separate assignment in the execution of the overall plan. These facts were finally recognized in the organizational parity which was granted to air power within our principal unified commands."[19] A unified national security establishment was required and Truman stated this would call for "new ways of thinking" in the defense establishment. Along with General Dwight D. Eisenhower, Truman determined that unified command was a necessity. In December 1946, Truman approved the first Unified Command Plan, creating unified commands in geographic areas, taking strategic direction from the Joint Chiefs.

General Eisenhower, returned from the "Crusade in Europe" and about to become Army Chief of Staff, strongly supported an independent Air Force, emphasizing that "no sane officer of any arm should contest that thinking."[20] The air commander and his staff, said Eisenhower, "are an organization coordinate with and coequal to the land forces and the Navy . . . that seems to me to be so logical from all of our experiences in this war—such an inescapable conclusion that I for one can't even entertain any longer any doubt as to its wisdom."[21] Moreover, stated Eisenhower, "the Army does not belong in the air, it belongs on the ground. Control of the Tactical Air Force means responsibility for the entire operating establishment required to support these planes. Assumption of this task by the Army would duplicate in great measure the primary and continuing responsibilities of the Air Force and, in effect, would result in creation of another air establishment."[22]

Despite the Navy's misgivings, support by Truman, Eisenhower, and the Congress resulted in the landmark legislation of the National Security Act of 1947. Establishment of the United States Air Force was but one part of the legislation which created the modern American national security establishment. James V. Forrestal, to become the first Secretary of Defense, described the Act as "one of the most far—reaching and important steps in Government organization since the founding of the nation itself."[23] The National Security Act is distinguished by its unitary approach, tackling national security as a single problem.

The Act gave the Army Air Forces independence, but it did not conform to what any of the services had originally wanted. Lieutenant General Ira C. Eaker noted that the Act "legitimized four military air forces."[24] The fact was that the architects— including Major General Lauris Norstad and Admiral Forrest Sherman—had to navigate through heavy shoals to reach a compromise. Stuart Symington, the first Secretary of the Air Force, observed: "A bill which was considered better could not have gotten everybody's approval; and therefore would not have given the President the opportunity to show agreement to the Congress and the people. I don't say this is a good book, but I do say it is a good chapter."[25] The National Security Act was a first step toward an integrated defense establishment. Its major weakness remained that the Secretary of National Defense operated as a coordinator (as the Navy had desired), not as an administrator, as the Army and Air Force wanted.

The relatively new U.S. defense establishment was tested by the Korean War, which had a major impact on the Air Force and shaped the American political military landscape for half a century. Its lessons for air organization appeared contradictory. Although during the war the Tactical Air Command was being restored as a major command, this was not the premier lesson of the war. Rather, this first major war of the nuclear era resulted primarily in a buildup of the Strategic Air Command (SAC) as the nation's nuclear deterrent and prompted the Eisenhower administration to initiate the New Look military policy.

President Eisenhower knew the flaws in the National Security Act only too well. A staunch enemy of service parochialism, Eisenhower was determined to reorganize the Defense Department. The 1958 Reorganization Act, a turning point in American military organization, removed the military departments from the operational chain of command, the Joint Chiefs becoming a conduit between the Secretary of Defense and the unified and specified commands. This reorganization gave the unified and specified commanders control over U.S. combatant forces. The air component commander would serve under the theater combatant commander. Responsibility for preparing and supporting forces remained with the military department.

Organization and control of the air war in Southeast Asia amounted to a conglomeration, an absence of unity of command of air power. General William W. Momyer, Seventh Air Force commander, reported to General William Westmoreland in Saigon on air operations in South Vietnam, and to General Hunter Harris, commander of

Pacific Air Forces in Hawaii, for air operations over North Vietnam. In addition, the Strategic Air Command, a specified command, received target assignments direct from Westmoreland, dropping more than a third of the bombs in South Vietnam. Consequently, the Korean War and the conflict in Southeast Asia pointed to the increasing importance of tactical air forces. The success of SAC as the nation's nuclear deterrent meant that tactical air elements would most likely be engaged in conflicts although in both Korea and Southeast Asia strategic air power was employed. Korea, Southeast Asia, and the first Gulf War posited a gray area between the tactical and strategic.

Ever since World War II and formation of the Air Force in 1947, the USAF had advocated a more unified and centralized defense establishment. The war in Southeast Asia increased the pressure to strengthen the role of the combatant commanders. As Air Force General David Jones, chairman of the JCS, observed: "We need to spend more time on our war fighting capabilities and less on intramural squabbles for resources." These efforts resulted in the Goldwater-Nichols Department of Defense Reorganization Act of 1986, which gave more power to the chairman of the Joint Chiefs and the combatant commanders. Goldwater-Nichols designated the chairman of the Joint Chiefs as the principal military advisor to the President and Secretary of Defense, responsible for overall strategic planning.

On the heels of Goldwater-Nichols, in the 1990s the issues of organization, control, and direction of the American military again took center stage, in the wake of the collapse of the Soviet Union, the end of the cold war, and the success of the first Gulf War. The experience of the Gulf War, with its blurring of the tactical and strategic air missions, resulted in a historic reorganization of major air commands, the first since the formation in March 1946 of the Strategic, Tactical and Air Defense Commands. The Air Force reorganization of mid-1992 combined most of SAC with the Tactical Air Command and part of Military Airlift Command, to form Air Combat Command with headquarters at Langley AFB, Virginia, forming a link to Frank Andrews command in the 1930s of GHQ Air Force at Langley as well as to the Air Combat Command of 1941.

With the end of the Cold War and emphasis on war-fighting and joint operations, in 1999 the Air Force moved to bring the service in line with the U.S. national strategy of selective engagement. Ten Aerospace Expeditionary Forces (AEFs) were formed to make the Air Force of the twenty-first century an Expeditionary Aerospace Force (EAF. The EAF concept can be traced back to Air Corps humanitarian operations during the interwar period and to Tactical Air Command's establishment [with the Nineteenth Air Force] in the early 1950s of the Composite Air Strike Force (CASF), a reaction to the lack of a quick response force at the start of the Korean War.[26] The EAF concept proved itself in the 2003 war with Iraq. As Air Force Chief of Staff General John Jumper emphasized: "For the first time in the history of the Air Force, we relied on the Air Expeditionary Force to present the full spectrum of our capabilities to combatant commanders around the world. It is the right war-fighting construct for our 21st century Air Force."[27]

These issues of organization, control, and employment ultimately flow from the democratic process. General Arnold possessed a fine sense of process in a democratic society, honed in the grim interwar years, and in World War II, which he saw as a testament to what the American people could accomplish. The danger zone of modern war," Arnold emphasized, "extends to the innermost parts of a nation." Consequently, he noted, "air power will always be the business of every American citizen. It is the American people who will decide whether this nation will continue to hold its air supremacy. In the final analysis, our air striking force belongs to those who come from the ranks of labor, management, the farms, the stores, the professions, the schools and colleges, and the legislative halls."[28]

As we look back over a century of what I call "constant threads" or connective tissue in organization, spanning the years of flight from dusty fields; the humanitarian operations and record-setting flights of the interwar years; the immense global operations of World War II; the Korean and Southeast Asian conflicts; the long years of the Cold War; operations in the Balkans; and the two wars with Iraq, one does not have to proclaim a revolution in military affairs to recognize the enormous change in the way air wars are fought. What is clear is the evolution not only in technology, but also in organization and control of air forces that enabled American air power to become dominant in today's world, to conduct traditional Air Force missions with heavier payloads and faster reaction time. American air power is now uncontested. As one historian observed, "the aerial arms race, a central facet of the last fifty years, is over."[29] However, perhaps a cautionary note is appropriate: After World War II, General Arnold stated that "the principles of yesterday no longer apply. We must think in terms of tomorrow. We must bear in mind that air power itself can become obsolete."[30]

The advances in organizing and controlling air war were made by men of faith who believed that evolving technology demanded control by airmen. World War II became the crucible that proved the airmen's point as well as the paramount need for an independent Air Force and unity of command. The last fifty years have seen development of the unified command system, a focus on strengthening joint warfighting ability, on tamping down what Eisenhower termed "parochialism."

The end of the Cold War and the stunning events since the collapse of the Soviet Union have again emphasized the need to determine how best to organize the American military to meet the challenges that lie ahead.

Notes

1. Woodford Agee Heflin, ed., *The United States Air Force Dictionary* (Princeton, N.J.: Van Nostrand, 1956), 34.

2. "Third Report of the Commanding General of the Army Air Forces to the Secretary of War," 12 Nov. 1945, *The War Reports of Marshall, Arnold, and King* (Philadelphia and New York: Lippincott, 1947), 455.

3. Carl A. Spaatz, "The Future of the Air Force," November 1923, Air Force Historical Research Agency [OL-A], Washington, D.C.

4. Testimony of Maj. Gen. Frank M. Andrews to Senate Appropriations Committee, April 1937.

5. Gen. Peyton C. March, "Lessons of World War I," *War Department Annual Reports, 1919* (Washington, D.C., 1920), 471–78.

6. Interview, Lt. Col. J. B. Green with Lt. Gen. Ira C. Eaker, 11 Feb. 1967, Washington, D.C.

7. James H. Doolittle dissent to Final Report of War Department Special Committee on Army Air Corps (Washington, D.C., 1934), Baker Board Report, RG 18, Modern Military Br., National Archives (NA).

8. Report of Browning Board, 7 Jan. 1936, AAG 334.7, Boards General, National Archives, cited in Chase C. Mooney, *Organization of Military Aeronautics, 1935–1945,* AAF Historical Study 45, Washington, D.C., 1945.

9. Letter, Gen. Jacob E. Smart to Herman S. Wolk, 18 Feb. 1983.

10. "First Report of the Commanding General of the Army Air Forces to the Secretary of War," 4 Jan. 1944, *War Reports of Marshall, Arnold, and King,* 304.

11. Lovett to Arnold, 14 Oct. 1942, RG 107, File 452.1 (7), Production Item 11A, NARA, cited in George M. Watson, Jr., *The Office of the Secretary of the Air Force, 1947–1965* (Washington, D.C.: Center for Air Force History, 1993), 21.

12. Arnold to Lovett, "Aircraft Production for 1943," 20 Oct. 1942, ibid., cited in Watson.

13. Ray S. Cline, *Washington Command Post: The Operations Division* (Washington, D.C.: Office of the Chief of Military History, Department of the Army, 1951), 249.

14. Summary of Patch-Simpson Board interview with Gen. Marshall, 5 Sept. 1945, Patch-Simpson Board Files, RG 165, Records of the War Department General and Special Staffs, MMB, and NA.

15. Arnold's comments and plan are published in Maj. Gen. Otto L. Nelson, Jr., *National Security and the General Staff* (Washington, D.C.: Infantry Journal Press, 1946), 337–42.

16. Arnold to Eaker, 19 June 1943, CM OUT 8090, cited in Cline, *Washington Command Post.*

17. Ibid., 253.

18. Cited in Gen. George C. Kenney, *General Kenney Reports: A Personal History of the Pacific War* (1949; Washington, D.C.: Office of Air Force History, 1987), 352.

19. Special Msg. to the Congress Recommending the Establishment of a Department of National Defense, 19 Dec. 1945, in *Public Papers of the Presidents of the United States: Harry S. Truman, 1945* (Washington, D.C.: GPO, 1961), 555.

20. Hearings before the Committee on Military Affairs, Senate, Department of Armed Forces and Military Security: Hearings on S.84 and S.1482, 367.

21. Eisenhower to ACS, G-1, et al., "Responsibilities of Staff Officers, Scope, Approach and Execution," 10 Dec. 1945, RG 165, ACS, Patch Simpson Board Minutes, box 922, MMB, NA.

22. Eisenhower to SecDef, "Tactical Air Support," 3 Nov. 1947, RG 18, AAG 322, box 104, MMB, NA.

23. Statement by James V. Forrestal to Senate Military Affairs Committee, 22 Oct. 1945.

24. Interview, Thomas A. Sturm and Herman S. Wolk with Lt. Gen. Ira C. Eaker, 17 Nov. 1972.

25. Symington to Webb, Bureau of the Budget, 3 Feb. 1947, C4-3 folder, Papers of James E. Webb, Harry S. Truman Library.

26. See Richard G. Davis, *Anatomy of a Reform: The Expeditionary Aerospace Force* (Washington, D.C.: Air Force History and Museums Program, 2003).

27. Gen. John Jumper, Chief's Sight Picture, "Resuming the AEF Battle Rhythm," 9 May 2003.

28. Gen. H. H. Arnold, "Third Report to the Secretary of War," November 1945, *War Reports of Marshall, Arnold, and King,* 468.

29. Gregg Easterbrook, "Week in Review," *New York Times,* 27 Apr. 2003, 5.

30. Henry Harley Arnold, *Global Mission* (New York: Harper, 1949), 615.

Forging American Air Power: Patrick, Arnold, and Doolittle

Dik Alan Daso

In the history of American military aviation, the making of heroes is generally associated with personal sacrifice or technological achievement. In the flying business, we remember these heroes in different ways. We name airfields after those who have fallen—Selfridge, Andrews, Keesler, Westover, and Nellis are but a few of these so memorialized. We erect statues and award medals to honor others for courage, service beyond the call of duty, for remarkable achievement—Doolittle, Risner, Sijan, Arnold, and Mitchell come to mind. These physical representations stand as constant reminders of obstacles overcome, sacrifices made, and milestones achieved.

During the formative years of U.S. Army aviation, it was often those who simply survived the hazards of early military flying operations that eventually made the most significant impact upon the U.S. Air Force (USAF) as we know it today. Many of these were laudable in their own right—Spaatz, LeMay, Hansell, and Vandenberg fall into this category. Yet, there are three who stand out in USAF culture for achievements during their military careers that are today taken for granted—Mason Patrick, Henry Arnold, and James Doolittle. Although these names are familiar ones, many of their most significant contributions to the evolution of the "Winged Crusade" are not only synergistic but remain unappreciated in the study of the history of the United States Air Force.

Mason Patrick's Contributions

Mason Mathews Patrick was born in December 1863, a son of a Confederate Civil War surgeon. He graduated West Point in 1886, second in his class behind John J. Pershing. Ranking so high in his class, he chose to enter the engineering branch of the Army—at that time, the most sought-after assignment. He was rapidly promoted to brigadier general prior to World War I and served as a troubleshooter for Pershing's expeditionary forces in Europe. Brilliant and efficient, Patrick was later called upon to command the Army's air arm. Rapid demobilization occurred and budgets vanished, leaving the airmen "running in circles."[1] In September 1921, Patrick took command of a disorganized Air Service. In his view, it was "a tangled mess." Although he accepted the job, he considered it "a most onerous duty," certainly one that had little positive side and massive hurdles to overcome.[2]

Legitimized the Air Service

Patrick was one of those in history who was blessed by perfect timing and compatible intellect. He played a vital transitional role in the evolution of American air power. He linked the *Ancien Régime* of the land-bound Army to the progressive elements of the new air-minded one. He maintained his belief in military discipline but tempered its application to fit the personality of an emergent "new breed of cat"—the military aviator. He had witnessed Army Air Service pilots in action during the Great War and had been impressed enough to write that "no finer body of young men ever went forth to do battle for their country. . . . The only thing comparable to it is the story of the stirring deeds of the knights of old who rode forth to redress the wrongs of the distressed."[3] This observation from one of the Army's elite—an engineer.

Patrick recognized early during his command that the culture of the Army aviator was something different than that of the Old Army. "I have never known any men who so persistently, insistently, 'talk shop' as do air pilots," he wrote, "and they have a language of their own which none but the initiated can understand."[4] So, he decided to get initiated. Patrick became the oldest Army man to earn his junior aviator wings. He did so during the summer of 1923 at Bolling Field, just outside Washington, D.C. He was almost sixty years old. Such an accomplishment was remarkable, but the purpose for achieving the rating—to better understand and earn the respect of his men—reflected true courage and wisdom. He understood that to command respect from airmen, he had to be one of them. He applied this philosophy during his tenure as chief of the Air Service.

Recognition of Air Doctrine

What factors came together during 1921–26 that allowed Patrick to achieve success in establishing a functional, even respected, Air Corps? First, the rapid development of aviation technology was something relatively new to Army leadership. Patrick, for example, had himself witnessed the capabilities of newly developed airborne radio sets: "I talked from the ground to a pilot in a plane a mile high, saw him obey the oral orders given him, and heard his comments perfectly."[5] His ideas about doctrinal application of air power were shaped by an understanding of aviation and the possibilities for its use.

More broadly, Patrick understood the true potential of military aviation. "It should be borne in mind," he wrote, "that an Air Force is highly mobile—it can move rapidly from place to place, cover great distances in comparatively short periods of time."[6] Experiments were ongoing in the field of transportation of airplane supplies by air, base construction at deployed locations, and a massive heavy bomber, the NBL-1 Barling, was built and flown near Dayton, Ohio.[7]

As tremendous as the potential for the application of the new air weapon might have seemed, there remained many hurdles to overcome before effective military capabilities could be realized. Scanty budgets were only surpassed by a maze of paperwork that made procurement and administration of the air arm nearly intolerable.

Patrick noticed that "the art of designing and building aircraft was changing and improving almost daily."[8] To many, continuous improvements seemed a tremendous advantage. In reality, rapidly changing technology was extremely costly because by the time a contract was awarded, the technology was already becoming obsolete. Even as aircraft were being built, there were others that could surpass the older design. This conundrum has ever since been a reality in the aerospace industry and military acquisition system. Patrick appreciated the technological aspects of aviation, but never had the budget or the support to make significant progress in airplane technology during his years in command. His appreciation for technology was itself a major step beyond the arcane attitudes of most of the Old Army. Despite the fact that most airmen already understood the changing technology problem, Patrick's acceptance by members of the Old Army paved the way for more significant steps toward the establishment of an air arm of some autonomy by 1926 with the establishment of the U.S. Army Air Corps.

One of the ways that Patrick kept the respect of the Old Army leadership was to impart discipline upon his men. His role in Billy Mitchell's court martial and the threat of one he himself issued to his own chief information officer, Henry Arnold, were supported by his superiors. His reputation as a disciplinarian, although disparaged by his aviator subordinates, assured that often dangerous stunting was reduced while well-planned aerial achievements stood a far greater chance of success.

Mason Patrick's unsung contribution to the "Winged Crusade" was the legitimization of air power within the Old Army. This was the first step in creating an environment, both doctrinally and politically, which would eventually recognize independence for the air arm.

Henry Arnold's Contributions

There is some irony in the eventual rise to command of the same young officer that Patrick had attempted to court martial in the late 1920s. Henry Harley Arnold, not known as "Hap" until 1931, had adopted much of Patrick's deliberate leadership style and resistance to immediate separation from the U.S. Army. His accomplishments as Army Air Forces (AAF) commanding general were remarkable, particularly from the administrative side. But perhaps his most important contributions to America's air arm were related to career-long efforts to improve air power technologies which today remain the backbone of expeditionary air forces around the globe.[9]

Mobility

Early flight was a fair weather undertaking. On 28 November 1911, as winter settled over the Washington area, Lieutenant Henry Arnold and the College Park aviators boxed up their planes and moved to Barnes farm, near Augusta, Georgia, hoping for more temperate flying weather. Unlike the planes of today, these fragile craft could not be left out in the open. They were constructed of cloth and wood, and even though they were finished with several coats of "dope"—a clear lacquer-type sealant and stiffener—the planes could not stand bad weather for any length of time. In the first ever aerial unit deployment, all of College Park's personnel—39 Signal Corps specialists plus the officers, their equipment which included motor vehicles, horses, mules, aircraft, and spare parts were loaded upon several train cars and transported to Georgia.[10]

The way that the Wright military planes at that time were transported was to remove the tail and skids and shove the assembled wings into a boxcar endwise. Then the tail assembly and the landing assembly could be fitted in the car as well so that the plane could be assembled very quickly after arriving at the destination.[11] From the very beginning, Arnold had been involved with deployment of airplanes. He recognized the logistical challenges involved with such deployment and dealt with them for the rest of his career.

Aerial Refueling

Strategic bombers must fly tremendous distances. To increase aircraft range for pursuit planes and bombers, Arnold approved the trials of a new, dangerous, potentially revolutionary advance in aviation operations—mid-air refueling. Initially, nothing more than hoses, ropes, and gas cans provided such capability. Audacity and fearlessness played a larger role in the success of the trials than the machinery involved. On 27 June 1923, 1st Lieutenant Frank Seifert and 1st Lieutenant Virgil Hine achieved two successful contacts in a modified DH-4 aircraft. A second, even more successful test occurred in August.[12]

Arnold had no doubt of the potential import of this event, still considered a stunt by most of the general public and even many aviators as well. Arnold was so pleased with the success of the refueling tests that he added it to the 1923 Rockwell Field "Holiday Greetings" letter:

> In performing the two aforementioned flights Rockwell Field presented to the world a new mode of replenishing gasoline and oil supply of an airplane while in flight. While the great benefits to be derived from refueling in the air are probably unappreciated at this time by many people in aviation circles, it can only be a matter of a few years until the pioneer refueling work done at this station will be the basis for operating airplanes on long cross-country flights whenever it is needed to carry great loads or carry materiel or personnel to greater distances than the capacity of gas and oil tanks will permit. . . . These things were done in spite of the handicaps under which we labored . . . such as decrease in personnel, limited appropriations and inadequate supplies during the year that has passed.[13]

Arnold's fascination with aerial refueling continued through World War II, using technology developed by the Royal Air Force (RAF). Basic tests were accomplished between modified B-17 and B-24 bombers, but the refueling system was so difficult to use quickly that it never went into production. Yet, Arnold pressed for results despite the inherent technological challenges of the operation. The fact that in-flight aerial refueling is the great enabler of all global air operations speaks much to Arnold's vision during these years.

Aerial Deployments

Another one of Arnold's significant achievements occurred while he was commander at March Field but did not take place there. Lieutenant Colonel "Hap" Arnold won his second Mackay trophy as commander of a flight of ten Martin B-10 bombers that flew round trip from Washington, D.C. to Fairbanks, Alaska. The first all-metal, low-wing, retractable gear monoplane, the B-10 bomber was the most technologically advanced airplane in the Air Corps inventory.

Planning was meticulous. A poor showing as a result of a poor plan would have been a catastrophic embarrassment, particularly since the Air Corps was still stinging from its performance while carrying the U.S. mail.[14] The trip was not Arnold's idea; it was most likely Assistant Chief of the Air Corps Oscar Westover's answer to the media's assault upon the sole source purchase of the bomber. Arnold arrived at Wright Field the last week in June to find planners hard at work on the mission. Assessing the progress and the possibilities for distribution of logistics and the readiness of the planes, he determined that the earliest mission departure date might be 10 July.[15] He informed Air Corps Chief Benjamin D. Foulois and Westover that adjustments to the proposed departure date were required and Arnold was called to Washington to discuss the particular objectives for the mission. During the ensuing meetings over the next two days, Arnold was given complete authority for the flight. He wrote: "I received about 4 sets of letters telling me that I was holding the sack with regards to safety, hazard, success, and risk! I in turn told them that I would not say when I would start on the flight until the planes were ready."[16] He immediately returned to Dayton to oversee the final preparations.

Unexpectedly problems arose with the supply lines to Fairbanks, the target destination. Striking longshoremen prevented commercial ships from moving supplies. Rail routes through Canada were available but could not transport the total amount of supplies needed. The U.S. Navy did not care to assist, and a transport ship normally reserved for soldiers was even considered as a possible alternative. Finally, the Army acquired a small barge, *El Aquario,* that carried half of the gasoline required for the Alaska portion of the journey. The rest went by rail.

By 6 July, six planes were ready for a shakedown flight. Arnold selected Major Hugh J. Knerr, the flight commander for the second three Alaska-bound B-10s, to lead it. At dinner that night, Arnold briefed Knerr on the requirements for the mission—a non-stop flight to Dallas and a non-stop return. Knerr was to test and record the gas and oil consumption of the planes so that flight planning could be refined for the flight to Alaska.[17] Arnold realized that a mission of this nature was dependent upon prepositioned supplies. This shakedown flight was specifically designed to refine the necessary logistics—critically restricted by striking longshoremen—so that only the absolute minimum tonnage would be shipped to Alaska. This type of operational planning was essential to smooth operations away from local operating facilities. The Dallas trip was an important part of the Alaska mission.[18]

The six aircraft would fly to Dallas on Saturday, 7 July, and return the following day. The planes actually flew to Dallas and back the same day. Arnold called it a "magnificent flight."[19] The success of the test flight reinforced Arnold's hopes. After the Dallas six-ship formation had returned, plans were finalized and the departure date was set for 18 July 1934. Foulois, initially disappointed in the date, realized that a perfect showing one week later was far more desirable than accidents caused by a rushed departure. For the next few days, Arnold had a few moments to relax. In his correspondence, he praised the enlisted men who worked all day and night to ready the aircraft: "They are a mighty fine bunch." Arnold also assigned the pilots and the crew chiefs to one specific plane. He believed that the crews would take a greater interest in their aircraft since they would be flying it.[20]

Just two days before departing, logistics concern befuddled Arnold once again:

> Well our military establishment most certainly breaks down when an extra load is imposed. Sometimes it is personnel and sometimes it is materiel—who knows which it will be in advance. . . . We have been delayed day after day due to mechanical defects and adjustments which—in my humble opinion—should have been corrected before we ever arrived here.[21]

The logistician in Arnold was disappointed in certain aspects of B-10 production, the supply officer in him pulled out what little hair remained trying to resolve fuel and consumables issues, and the pilot in him was raring to get going. His experiences served only to reinforce that which he already knew too well: without superior production and appropriate supply, there was no true air power. He remembered that lesson as commanding general during World War II.

On Tuesday, 18 July, the 10 Martin B-10s departed Wright Field at 7:00 a.m. Arnold's mission had three specific purposes: first, to test the newest Air Corps bombers (to include photo capability); second, to determine the practicability of moving a tactical squadron from the continental United States to Alaska; and third, to demonstrate our good will to the people of Alaska (which was not yet a state).[22]

Arnold's team departed Washington the following morning; their first stop, Wright Field. The next leg, two aircraft developed engine trouble and returned to Dayton. After quick repairs, they caught up with the others at Minneapolis. The remainder of the journey to Fairbanks was marked by smooth airplane performance and warm welcomes at every landing place. On 22 July, Arnold decided to lay-over one day so that the mechanics could accomplish required periodic checks and maintenance on the bombers. "This day of rest was extremely beneficial to the airplanes," Arnold explained to Bee.[23]

The publicity portion of the mission tried Arnold's patience. "I am getting good willed to death," he complained. He could not wait to reach the wilderness areas of Fairbanks. "When we get out of the entertainment area we will all be much better off. The people are all wonderfully hospitable, too much so but I guess that is all in the game."[24]

Even while airborne, the travelers could not escape the good will mission. Halfway between Prince George in British Columbia and Whitehorse in the Yukon, Arnold received a radio message from the governor of Alaska who offered an official welcome. The radios that had been installed at Wright Field worked tremendously. According to Arnold, they had a "wonderfully big range."[25] In fact, this Alaska flight was the first Air Corps long-distance mission that maintained continual contact with ground stations along their entire route of flight, even deep into Canada—a tribute to the equipment and the logistic planning.[26] Moreover, the B-10s made the journey from Whitehorse to Fairbanks in four and one-half hours. The same journey took two weeks by boat and one month overland. The harsh terrain and lack of improved roads only emphasized the practical aspects of air travel in the Yukon.[27]

The Alaska flight rectified a poor public view of the Air Corps which had resulted from the difficult circumstances of assuming responsibility for delivery of the air mail in early 1934. It also demonstrated the long-range capability of the B-10 and those support elements that were required to accomplish such missions. During these years, Arnold began to understand the very nature of American air power. To optimize its effectiveness, he devised a philosophy which would enable the buildup of forces at the quickest, most efficient pace possible—it was all about balance.

Arnold's Doctrine, a Balanced Air Approach, 1938–1945

Even before the March 1939 restructuring, Arnold had chaired a 1936 committee examining how best to create a "Balanced Air Program." There was nothing unusual in his report; in fact, it followed very closely the recommendations made previously by the Drum Board in July 1934. The numbers reflected in each report for personnel and planes were similar. Surprising today but realistic at that time, the forecast for airplanes totaled 1,399 in 1936, increasing to a meager 2,708 in 1941.[28] Arnold's report primarily attempted to reckon with recovering budgets; no mention was made of scientific research or technological development. Rather, the program's primary concern was to save dollars in all areas except purchasing airplanes.

Arnold considered it amazing that, despite the rapidly changing global situation, the War Department had done little "clear-cut thinking about American air power." Mussolini's invasion of Ethiopia in 1935, the unveiling of the *Luftwaffe* in the Spanish Civil War, Adolf Hitler's remilitarization of the Rhineland in 1936, and the Japanese bombing of Shanghai in 1937 all seemed signals of instability and potential future conflict.[29] These events had an impact on Arnold.

In September 1937, while addressing the Western Aviation Planning Conference, Arnold summarized his philosophy for creating an aeronautical institution in America second to none:

> Remember that the seed comes first; if you are to reap a harvest of aeronautical development, you must plant the seed called *experimental research*. Install aeronautical branches in your universities; encourage your young men to take up aeronautical engineering. It is a new field but it is likely to prove a very productive one indeed. Spend all the funds you can possibly make available on experimentation and research. Next, do not visualize aviation merely as a collection of airplanes. It is broad and far reaching. It combines manufacture, schools, transportation, airdrome, building and management, air munitions and armaments, metallurgy, mills and mines, finance and banking, and finally, public security—national defense.[30]

Arnold's statement described the evolving technological system of air power in broad terms, even if he did not make a distinction between empirical and theoretical research. If the Air Corps had little money for R&D, then perhaps universities and industry could be persuaded to find some. After all, it had been the Guggenheim Fund that had fostered aeronautical departments at several universities almost a decade earlier.[31]

No matter the source, experimental research appeared the key to future air power. Arnold had very cleverly linked Air Corps development to civilian prosperity in the aviation industry, hoping that civilian institutions would pick up research while the Air Corps acquired planes. His ideas reflected the "Millikan philosophy," that of bringing the center of aeronautical science in America to Caltech, which had shaped that university since the 1920s. This philosophy, coupled with Arnold's realization that air power was a complex system of logistics, procurement, ground support bases, and operations, guided his vision for future growth.[32] Arnold's approach to air power development demonstrated what today is commonly known as the Military-Industrial-Academic complex.[33]

Since August 1936, Arnold had advocated a "Balanced Air Program"—one that emphasized not equilibrium in aircraft types (fighters versus bombers)—but one that supported balance between airplanes, personnel, and bases. Changes in one of these elements necessarily affected the other two, and without balance between them, inefficiency and budgetary waste would result. In effect, Arnold believed that a mathematical formula of sorts existed that maximized the utility of air power. When one part of the formula was lacking, emphasis was logically placed in that area until the formula was once again balanced. The buildup of air power during his tenure was predicated upon an ever-more complex equation of combat crews, airplanes, and base facilities growing in harmony so as to maximize the potential of available air power. It was this basic plan that influenced President Franklin D. Roosevelt's expansion of the Air Corps in November 1939.[34] Arnold's increased emphasis on R&D was intended to improve the capability of Air Corps aircraft early in the expansion program.

In Arnold's 1936 report, he noted that "an acute shortage exists in the number of primary training airplanes to train the number of pilots to provide for a balanced program."[35] Logically, Arnold placed immediate emphasis on solving the training issue and after General Headquarters (GHQ) was subordinated to the Commander of Air Corps in March 1939, Arnold had the authority do things his way—a situation that did not always please the new GHQ Commander, Delos Emmons. "Tooey" Spaatz, who had recently changed the spelling of his name to ensure the proper pronunciation ("spots"), understood Arnold's methods.

Unmanned and Precision Weapons

During World War I, Arnold had been assigned to the War Department as a staff officer. While on the staff, he became involved in a very secret project that involved a contraption known as the Liberty Eagle. This small, unmanned biplane was designed to be launched toward enemy territory and then plunge to the ground and explode in the midst of enemy targets. Although never used in combat, the "FB" (short for flying bomb) was produced in small numbers. More importantly, the program had interested Arnold. He would later try to reactivate the project as World War II was beginning, but the limited range of the device resulted in its cancellation. Another project, however, was actively pursued and even tried in combat over Europe—the precision control of glide bombs and war-weary aircraft.

That project was known as Aphrodite. Two different types of weapons, a series of glide bombs (GB), and remote controlled, and nitro-starch filled, war-weary bombers remotely piloted from a trailing mother ship were used during this operation. Aphrodite was an attempt to protect airmen from deadly flak belts overhead high-value targets. The glide bombs, capable of gliding one mile for each thousand feet altitude, were initially unguided and not very accurate with a computed error of up to one mile around the target. Bombers using radio-controlled steering and aimed by using a television camera mounted on the bomb to direct later models. This method, of course, was dependent upon visual conditions if a pinpoint strike was desired. These standard one or two thousand pound bombs affixed with a small set of wings were employed by the Eighth Air Force from mid-1943 until May 1944.

Immediately after the GB series was shelved due to unacceptable inaccuracy, the radio-controlled "Weary Willy" bomber took to the skies. In essence, each orphan B-17 or PB4Y (a naval B-24) was packed full with 20,000 pounds of TNT (nitro-starch). The pilot took off, flew the plane to the English Channel, the bombardier armed the explosives, and the two crewmen parachuted out of a large hole cut in the bottom of the bomber's fuselage. These monstrous "bombs" were originally planned for use against submarine pens and V-1 and V-2 missile sites. A few blew themselves out of the sky when the explosives went off prematurely. About a dozen actually were guided across the Channel and detonated near targets in France, inflicting little damage but creating huge holes in the ground. Aphrodite, a concept well ahead of the technology of the day, was eventually shelved when the British began to fear reprisals by V-2 rockets as a result of errant Aphrodite attacks.[36]

Clearly, the purpose of several of these projects was to deliver weapons to targets while increasing the survival potential for airmen. Successful, accurate employment of such "stand-off" weapons would not be possible for nearly five decades.[37] Technology had not then caught up with physical capability of existing weaponry. Yet, Arnold's belief in science was unyielding.

Lieutenant General James H. "Jimmy" Doolittle, who chaired the Air Force Scientific Advisory Board from 1955 to 1958, offered this unique observation about "Hap" Arnold:

> General Arnold was unique in his ability to anticipate and prepare for the future, and when he got together with Theodore von Kármán it was a very fortunate thing indeed, because while General Arnold was not a highly technical man he did understand the importance of science and technology, and while Dr. von Kármán was not strictly a military man he realized the importance to the military of the mobilization of science.[38]

Although their individual and joint contributions were vital, it was the actions which resulted from the collaboration between these men and the impact which those actions had on the evolution of the technological system of air power which first began molding the fledgling Army Air Force in 1945 and has continued to shape the independent Air Force today.

General "Hap" Arnold's impact on today's Air Force cannot be understated. Aside from his accomplishments as the organizer of the Army Air Forces during World War II, including more than 2.4 million personnel and nearly 300,000 aircraft, his vision for a technologically based Air Force remained steadfast. Throughout his career, he envisioned a mobile air force and practiced ways to make it so. He envisioned aerial gas stations that could keep his aircraft flying around-the-clock and could then deploy to any location on the globe whenever needed. Arnold understood that to optimize air power assets a balanced approach to training, production, and basing was essential so as not to waste resources—both time and money. He supported research and development to enhance design and performance of his aircraft. Further, he insisted upon some forward thinking concerning weapons. Specifically, Arnold wanted to keep his aircrews out of deadly flak belts while still allowing them to apply precise and deadly force on the battlefield. Today's Air Force and all of its assets—capable of global engagement and rapid, accurate employment of deadly force or humanitarian assistance—is a direct descendent of Arnold's vision.

There seems an interesting connection between the three subjects of this article. As Patrick had disciplined Arnold once as a young officer, Arnold had once grounded Doolittle for "stunting" at Rockwell Field, California. Perhaps there is something to be said for an independent spirit and its impact on the evolution of American air power.[39]

James Doolittle's Contributions

James Harold "Jimmy" Doolittle was accepted into the graduate school at the Massachusetts Institute of Technology (MIT) in the fall of 1923. Jimmy moved to Cambridge but remained assigned to McCook Field to retain his flying skills. This allowed for the unusual opportunity to study aeronautics in the classroom and then test academic hypotheses in the air at McCook. After several months of classes, Doolittle selected a topic for his thesis project. He planned to study accelerations, or "G" forces, and their impact on aircraft and pilots in flight

Aircrew Physiology

After over 100 flight hours in a Fokker PW-7, he terminated his research after nearly ripping the wings completely off of the aircraft. Doolittle concluded that the increased pressure altered the mechanical components of force exerted upon the wing. These changing forces determine where and when the wing might suffer catastrophic failure. He concluded that the manufacturer's design limits were quite close to actual failure acceleration tolerances, but further experimentation had to be cancelled due to the failure of the Fokker's wings.

More importantly, Doolittle determined the basic rules for effects of "Gs" on pilots and that short duration acceleration had less effect than sustained acceleration. He also determined that sustained "G" forces resulted in physical impairment and also noticed that higher blood pressure in pilots meant that they could sustain high-G acceleration for a longer time. This was such important work that Doolittle was awarded his Master's degree in 1924 and another Distinguished Flying Cross, but not until 1929.[40] These documented discoveries are, today, a distinct element of basic military pilot training and critical in piloting modern high-performance military aircraft.

Effects of Wind on Flying Performance

Since the Army had agreed to two years' academic time at MIT, Doolittle simply continued the search for a dissertation project. Doolittle liked school and saw an opportunity to "make some lasting contribution to aeronautics through research."[41] He decided to attack a particularly disputed subject among experienced pilots—the effects of wind on airplane performance.

After many trial flights, Doolittle determined that wind affected ground speed and thereby impacted climb gradients. But his professors rejected what appeared only to be supposition. They required support with mathematical calculations. He used formulae from the famous Ludwig Prandtl, professor of aeronautics from Aachen, Germany. He took photographic data of time and distance during flight-testing and measured as many variables as possible.

Doolittle's research described the impact of wind on ground speed of aircraft. Understanding this concept is still a fundamental concept learned by all modern pilots in basic flight school. After including the appropriate arithmetic calculations in his report, Doolittle was awarded a Doctor of Science degree in June 1925.[42] It would be only a few years later that Doolittle found an occasion to apply his research to practical application in the air.

Blind Landing

In August 1928, Harry F. Guggenheim announced that funding was being made available to open a "Full Flight Laboratory" for the express purpose of studying flight in poor weather conditions. The Guggenheims had requested the use of a skilled pilot to act as the director of their newest attempt to consider "fundamental aeronautical and aerodynamical [sic] problems." Five flyers, both Army and Navy, were nominated and Doolittle was selected. Vice president of the Guggenheim fund, Captain Emory S. Land, an officer in the Construction Corps of the Navy, had selected Doolittle because of "not only a lifetime of flying, but a technical education that has given him a distinct advantage in the development of new equipment."[43] No other pilot could boast such credentials.

While the technicians studied the instrument problem, Jimmy researched previous weather-flying programs and found that a majority suggested physical modifications to aircraft in attempts to better sense when the ground was approaching. This meant adding a device that actually made contact with the ground some time before landing would occur. These physical modification were deemed too intrusive to the critical nature of the landing and, therefore, impractical.[44] Essentially, the Guggenheim team had to begin from scratch. The chosen method centered on the development of new, highly accurate flight-performance instruments.

The result of their efforts was the invention of three new instruments. Kollsman's detailed craftsmanship and near-watchmaker precision resulted in the construction of an altimeter that was accurate to a known elevation plus or minus five feet. This improvement was ten times more accurate than anything then used in an aircraft. The Sperry team devised a gyro-driven artificial horizon that displayed aircraft bank relative to the surface of the earth and a directional gyrocompass that was set to a known heading and then maintained that heading without magnetic input but rather by the spinning of a gyroscope. The development of these three instruments paved the way for Doolittle's ability to takeoff, fly, and land without reference to the ground.

On the morning of 24 September, thick fog coated the Mitchel Field landing zone. Doolittle hopped in the NY-2 alone after having the beacons turned on and climbed into the fog. Joe Doolittle, Elmer Sperry, Jr., and aircraft mechanic Jack Dalton waited for his reappearance from the east side of the field. Success! But this was only a warm-up as the project sponsor, Harry Guggenheim, had not yet arrived from his nearby residence; the fog was too dense to drive quickly.

Doolittle was confident in what was about to happen. Guggenheim arrived and after a brief discussion it was decided that Doolittle would fly while covered by the hood and Ben Kelsey would sit in the front seat as a safety pilot and observer. The fog was still thick but had lifted so that the ground was becoming visible from the air. The restriction to his sight was absolutely essential to prove beyond doubt the accomplishment of Guggenheim's team.

Ben Kelsey placed his arms visibly on the outside of his cockpit, demonstrating that he was not in control of the airplane. Doolittle lined up for takeoff using his localizer reed device and off they went. He climbed straight ahead to 1,000 feet above the ground. After reaching that altitude, he turned 180 degrees to the left and pointed the plane back toward the airfield, just a little bit south of the landing zone. After passing the field, he initiated his final turn to the left and intercepted the "beam." He began his descent to 200 feet and waited until the vibrating reeds quit moving, indicating he was over the final descent point.

There was no device that told Jimmy if he was flying a proper glide angle for landing. A useable glide-path indicator would not be invented until a few years later. To accomplish a controlled landing from a known point, Doolittle had perfected a method for flying a known path by maintaining a known airspeed and correcting slightly for the direction of the surface winds. His dissertation had exposed him to the knowledge of winds that he required to compute and execute this final landing approach. He set his power to maintain 60 mph in a slight descent until he flew the aircraft into the ground. The strengthened struts absorbed the blow (about 12 feet per second) and, on most landings, resulted in a relatively smooth touchdown. A mark placed next to his throttle assured the proper power setting for the descent. On this historic attempt, however, he misjudged his descent rate and hit with enough force to bounce once back into the air, and then land for good. "Despite previous practice," he recalled, "the final approach and landing were sloppy."[16] Sloppy or not, the feat had been accomplished.

Doolittle's pride ran deep. For the rest of his life he believed that "assisting in the instrumentation necessary to do the flight was my greatest contribution to aviation."[46] The fact that Doolittle held this achievement above his aerial demonstration and racing accomplishments, above his still-to-come pioneering work with Shell Petroleum, even above his successful execution of the Tokyo Raid in 1942, and his command experiences in World War II is a clear indication that he understood the significance of this first-ever blind landing. Following Doolittle's blind landing, every aircraft equipped with the capability to fly in limited visibility was afforded the instrumentation needed because of the success of the Guggenheim Full Flight Laboratory's quest to solve the problems associated with blind flying and landing. Doolittle later recalled that "Flying by instruments soon outgrew the early experimental phase. It became a practical reality, and aviation entered a new era."[47]

High-Octane Fuel

Jimmy Doolittle realized that all facets of aviation technology, not just instrumentation, were undergoing rapid development, particularly engines. Larger, more powerful motors were guzzling aviation fuel by the millions of gallons. But fuel development had stagnated. Interestingly, there was a general understanding among pilots that West Coast gasoline provided better engine performance and more race victories. In fact, California gas was usually imported for racing planes. The high concentration of certain elements seemed to preserve pistons during races. Clearly, research was needed to find out why.

Because of his continuing association with the Army Air Corps as a reserve major, Doolittle was aware of new military planes under development that would require fuel of superior quality. High-octane ratings, approaching 100, were anticipated for these aircraft engines, which did not yet exist. Military planes in those days used a variety of fuels that ranged from 65 to 87 octane. Doolittle convinced his bosses that the research needed to develop 100-octane fuel would be an important investment in the future of the company:

> I believed that there was a future for 100-octane fuel, but there were no engines to use it, and it was like the hen and the egg, which came first? If you waited for engines to be developed for 100-octane fuel, they'd never be developed, because there'd be no fuel for them. So I felt that we had to go ahead . . . and make the fuel available. And when the fuel was available at a reasonable price, I felt that there would be enough engines developed . . . that it would be a profitable venture. This turned out to be true, but it was a gamble.[48]

During these years of economic turmoil, this idea was a hard sell. Ironically, Doolittle found that his academic credentials, not his pilot skills, made the greatest difference in convincing skeptics of the potential of 100-octane fuel. Jimmy believed that to sell technical ideas to technical people, the salesman needed excellent technical credentials. An MIT Doctor of Science degree was the ticket that gained him audience with those scientists who made company decisions. "I think there are two great advantages to an advanced degree," Doolittle said, "one is the increased knowledge and greater capability that you have, and the other is the prestige it gives you with your associates—particularly those who have advanced degrees."[49]

By 1934, Shell had produced its first batch of 100-octane fuel for testing at Wright Field, near old McCook Field, home of Army Air Corps flight-testing. The new fuel not only increased power, it saved between 10 and 20 percent on consumption. By 1938, the Air Corps was recommending 100-octane fuel for all of its combat aircraft.

The fuels used by the air forces gave engines at least 20 percent more power while planes developed 6 percent more speed and better climb performance than with lesser octane fuels. The Navy had also made the transition to high-octane fuels by 1938. The RAF was able, using U.S. 100 octane fuel, to get 1,700 horsepower from Merlin engines as opposed to 1,000 horsepower provided on lower octane fuel.

Commercial airlines also benefited from 100-octane fuel. Using the military fuel, they were able to shorten takeoff distance and increase range. Flying became more economical. It was not until 1942 that the distinction of octane change with changes of engine power became fully appreciated. The 80/87 and 100/130 numbers of today's fuels reflect this discovery. High-octane fuels allowed engines to use less fuel for more economical operation and longer range with no increase in temperatures. Water injection allowed even more power over the short term.[50]

The Tokyo Raid

Whenever Jimmy Doolittle is mentioned in a military history context, the 1942 Tokyo Raid must always be mentioned—particularly in the context of this article. Not only did Doolittle plan and lead the mission that struck at the heart of Japan on 18 April 1942, but he would do this in the face of rapidly changing circumstances and challenging conditions. What changed after the Doolittle Raid was America's spirit— from the depths of defeat to joyous success literally overnight. The Raid also influenced the Japanese military establishment which resulted in accelerated plans to attack Midway Island—a cataclysmic mistake for the Japanese fleet.

Several attack plans had been studied and re-studied. Doolittle's original concept was for a daylight raid after a night launch from the carrier. Naval planners, however, considered the problems associated with night operations, too risky and that plan was scrapped. The second plan consisted of a dawn departure, daylight raid, and then landing at dusk at Chinese airfields. This idea was shelved as fears over daylight detection mounted among the planners. The plan finally agreed upon between the Army and the Navy included a near-dusk take off and night raid on Tokyo. This option, it was thought, stood the best chance of achieving surprise under the "greater security of a night attack."[51] The plan depended upon a fast carrier run-in at night to get as close to the mainland as possible just prior to launch and then an immediate turn back toward Hawaii and a run for waters beyond the range of Japanese land-based aircraft.

The near-dusk plan had been splintered by the Japanese detection of the American fleet. This resulted in a daylight launch and jeopardized the possibility for a safe landing. Compass drift and winds had pushed most of the aircraft well off of their planned course. Unknown to the Raiders, the aircraft that had carried the homing radio beacons for the landing fields in China had crashed, and took with it any chance of finding the strips at night and in bad weather. Lastly, but fortunately, the original targets, planned for night recognition and attack, were large industrial zones and hitting at least some part of the complex would be much easier during broad daylight.

The question of American bombing doctrine must here be raised, and certainly was by Jimmy Doolittle. The final plans for this raid centered upon a night attack. American bombardment doctrine with which the AAF entered the air war was one of precision, high-altitude, daylight bombing. None of these precepts were followed in the actual planning or in the subsequent execution of the Tokyo Raid. Why?

This raid was not intended to do maximum damage; rather, it was intended to create a spectacle.[52] The attack was designed so that the Japanese people would clearly know that a foreign enemy had bombed Tokyo. The fires Doolittle had planned to set were to serve not only as beacons to the other fifteen B-25s, but also to dramatically, and undeniably, announce that the capital city had been bombed. The plan to spread the attack force over a 50-mile front, at night, would have created the effect of many, perhaps hundreds of aircraft, attacking the island. Further, the order forbidding the bombardment of the radio towers near Tokyo revealed that immediate dissemination of the news by Japanese radio was expected. This was exactly what President Roosevelt had in mind when he ordered the attack.

It was not doctrine, but flexible, creative thinking that generated such an audacious plan. In reality, the Doolittle Raid violated almost every accepted doctrinal idea for bombardment openly held by the AAF. The mission objective, in this case a psychological one, dictated the method. Doolittle, later as commander of the Eighth Air Force in Europe, remembered that lesson when he released his fighters from bomber escort to hunt down and destroy the *Luftwaffe* in the air or an the ground.

Doolittle, a famous racing pilot, holder of the first-ever awarded aeronautics-related Ph.D. from MIT, and Shell Oil vice president, has had tremendous impact on American aviation through education and experimentation. Today's flight crews all are taught how to handle "G" forces in physiology class and calculate the impact of wind in almost every aspect of active flight operations. His ability to convince the Shell Oil Company to produce 100-octane aviation fuels may well have been a contributing factor in the RAF's victory during the Battle of Britain in 1940. He was daring in combat as well as a test pilot, but calculated the risks before taking chances with his or other's lives. His personal leadership during the raid over Tokyo provided a morale boost when America needed one most.

When the lives of all three of these men—Patrick, Arnold, and Doolittle—are taken together over the wash of time, the resulting combination of each man's contributions and their ability to act upon them have helped to forge the foundation of the most powerful and responsive military force in history. Whether refueling in the air or developing newer fuels, whether landing through dense fog or bombing through a sandstorm, or whether delivering food to a distant disaster location, the historic foundations of today's globally mobile Air Force may be found in the lives of those pioneers that survived the early days of military flight and quietly forged a path to modern air dominance.

Notes

1. Mason M. Patrick, *The United States in the Air* (Garden City, N.Y.: Doubleday, 1928), 6–7. For a recent biographical approach to Patrick, see Robert P. White, *Mason Patrick and the Fight for Air Service Independence* (Washington, D.C.: Smithsonian Institution Press, 2001).

2. Patrick, 82–83.

3. Ibid., 44–45.

4. Ibid., 111.

5. Ibid., 33.

6. Ibid., 106.

7. The NBL-1 (Night Bombardment Long-distance) was named after its designer, Walter Barling.

8. Patrick, 94–95.

9. Much of the material in the Arnold section of this paper is revised from my previous work, *Hap Arnold and the Evolution of American Airpower* (Washington, D.C.: Smithsonian Institution Press, 2001). Used with permission.

10. Charles DeForest Chandler and Frank P. Lahm, *How Our Army Grew Wings* (New York: Ronald, 1943), 211; Brig. Gen. T. DeWitt Milling, "Early Flying Experiences," *Air Power Historian* 3 (1956).

11. Thomas Milling, Columbia University Oral History Interview, 32, National Air and Space Museum Archives, Washington, D.C.

12. H. H. Arnold, "History of Rockwell Field," 112–14; also Mauer Mauer, *Aviation in the U.S. Army, 1919–1939* (Washington, D.C.: Office of Air Force History, 1987), 183.

13. Henry H. Arnold (HHA) Memo to Rockwell Air Intermediate Depot personnel, 21 Dec. 1923, Robert Arnold Collection (RAC).

14. H. H. Arnold, *Global Mission* (draft), 48–51½; For some unknown reason, Arnold allowed an inexperienced B-10 pilot to take one of the birds out on a flight. The pilot ended up in Cook's Bay and the B-10 was swamped in 20–40 feet of icy water. Remarkably, the other crews were able to save the plane and drain the water from the fuselage. After over one full week of work on the airplane, it cranked up on the first try and flew the rest of the way to Washington, much to Arnold's relief.

15. HHA to Eleanor P. Arnold (EPA), 28 June 1934, Wright Field, RAC.

16. HHA to EPA, 26 June 1934, ibid.

17. HHA to EPA, 6 July 1934, ibid.

18. Hugh Knerr, "Washington to Alaska and Back: Memories of the U.S. Air Corps Test Flight," *Aerospace Historian* (March 1972), 20–21.

19. HHA to EPA, 10 July 1934, Wright Field, RAC.

20. HHA to EPA, 15 July 1934, ibid. Arnold championed what the modern Air Force calls the "dedicated crew chief" program.

21. HHA to EPA, 16 July 1934, Wright Field, RAC.

22. Arnold's Alaska Diary, 18 July–9 Sept. 1934, RAC.

23. Ibid.; HHA to EPA, 21 and 22 July 1934, Edmonton, Alberta, RAC.

24. HHA to EPA, 22 July 1934, Edmonton, Alberta, RAC.

25. HHA to EPA, 23 July 1934, Whitehorse, Y.T., RAC.

26. Ibid.

27. HHA to EPA, 24 July 1934, Fairbanks, Alaska, RAC.

28. Report of Special Board Appointed to Make up a Balanced Air Program, 5 Aug. 1936, 145.93-96, USAF/Historical Research Agency (HRA), Maxwell AFB, Ala.; also see Herman S. Wolk, *Planning and Organizing the Postwar Air Force: 1943–1947* (Washington, D.C.: Government Reprint Press, 2001), 12–20.

29. Arnold , *Global Mission* (draft), 56½.

30. Address of Brigadier General H. H. Arnold, Assistant Chief of the Air Corps, at the Western Aviation Planning Conference, 23 Sept. 1937, 168.3952-119, USAF/HRA. Emphasis in the original. This belief in research may have been the result of earlier association with Dr. Robert Millikan. In 1934, Millikan had warned military officials through the executive Scientific Advisory Board, established in the summer of 1933, that "research is a peace-time thing and . . . moves too slowly to be done after you get into trouble." Quoted in Michael S. Sherry, *Preparing for the Next War: American Plans for Postwar Defense, 1941–45* (New Haven, Conn.: Yale University Press, 1977), 123.

31. Richard P. Hallion, *Legacy of Flight: The Guggenheim Contribution to American Aviation* (Seattle: University of Washington Press, 1977), summarizes the entire story of the Guggenheim influence on the early years of American aviation.

32. In another speech, "Air Lessons from Current Wars," Address before the Bond Club, Philadelphia, Pa., 25 Mar. 1938, Arnold emphasized the foundations of air power as not just planes but also "the number of flyers, mechanics, and skilled artisans available . . . and the size and character of the ground establishments we lump under the general name 'air bases.'" Papers of Ira C. Eaker, box 58, Arnold Speeches, Library of Congress, Washington, D.C.

33. Michael S. Sherry, *The Rise of American Air Power: The Creation of Armageddon* (New Haven, Conn.: Yale University Press, 1987), 200–201.

34. James P. Tate, *The Army and Its Air Corps: Army Policy Toward Aviation, 1919–1941* (Maxwell AFB, Ala.: Air University Press, 1998), 170–71.

35. H. H. Arnold, "Report of Special Board Appointed to Make up a Balanced Air Program, 5 August 1936" (Confidential), 145.93-96, USAF/HRA.

36. Dik Alan Daso, *Hap Arnold and the Evolution of American Airpower* (Washington, D.C.: Smithsonian Institution Press, 2000), 185–88; also see Wesley Frank Craven and James Lea Cate, eds., *The Army Air Forces in World War II,* 7 vols. (Chicago; University of Chicago Press, 1955), 3:530–32.

37. During the Persian Gulf War in 1991, American air forces employed precision-guided weapons that had their foundations in this type of experimentation during World War II.

38. James H. Doolittle, Oral History Interview, 21 Apr. 1969, USAF Academy.

39. Much of the material in the Doolittle section of this paper is revised from my previous work, *Doolittle: Aerospace Visionary* (Washington, D.C.: Potomac, 2003). Used with permission.

40. James H. Doolittle, "Wing Loads as Determined by the Accelerometer" (Masters thesis, MIT, 1924), Doolittle biographical file, CD 608500-1, National Air and Space Museum Library (NASML), Washington, D.C.; General James H. "Jimmy" Doolittle, with Carroll V. Glines, *I Could Never Be So Lucky Again: An Autobiography* (New York: Bantam, 1991), 92–93.

41. Doolittle, *I Could Never Be So Lucky Again,* 95.

42. James H. Doolittle, "The Effect of the Wind Velocity Gradient on Airplane Performance" (Dr. of Sc. dissertation, MIT, 1925), CD 608500-4, NASML. Doolittle called it a Doctor of Science in Aeronautical Engineering, while other sources called it a Doctor of Aeronautical Science. It really did not matter; his was one of the first degrees of its kind ever awarded by any American institution.

43. Hallion, 111; see also Doolittle, *I Could Never Be So Lucky Again,* 135; James H. Doolittle, USAF Oral History Program, 20 July 1967, K239.0512-998, 5–7, USAF/HRA.

44. Doolittle, *I Could Never Be So Lucky Again,* 136–37.

45. Quoted in Hallion, 123; see also Doolittle, "Early Blind Flying," for throttle details.

46. James H. Doolittle interview transcript, n.d., Doolittle biographical file, CD 608500-3, NASML; see also Doolittle, interviewed by Leish, April 1960, 146.34-39, 15–16, USAF/HRA.

47. Rebecca Hancock Cameron, *Training to Fly: Military Flight Training, 1907–1945* (Washington, D.C.: Air Force History and Museums Program, 1999), 264–67; Doolittle, *I Could Never Be So Lucky Again,* 152.

48. Doolittle interview, 17, NASML.

49. Ibid., 18.

50. The importance of 100-octane is summarized in Carroll V. Glines, *Jimmy Doolittle: Master of the Calculated Risk* (New York: Van Nostrand Reinhold, 1980), 176–77.

51. Doolittle, "Report to CGAAF," 14.

52. James H. Doolittle, interview with Murray Green, 22 Dec. 1977, Los Angeles, Hap Arnold (Murray Green Collection), 13, USAF Academy Special Collections.

An American Way of War: The Quest for High-Altitude Daylight Strategic Bombing

Tami Davis Biddle

If one were to identify the dominant theme in the history of American strategic aviation, it surely would be the quest to perfect "precision" bombing. Having the ability to identify and hit specific targets was, from the start, central to the American ideology of long-range bombardment as a tool of warfare. Targets selected for attack would be those most integral and fundamental to the enemy war effort; their destruction would disrupt and confound an enemy's ability to make war, and undermine its will to make war. This theory, which was embraced in the midst of World War I and then refined during the interwar years, was attractive to Americans for a variety of reasons. First, it aligned itself with intuitive ideas about the nature of industrial-age warfare and the ability of modern states to wage war. Second, it resonated with an American instinct for the application of technology to complex problems, and with the emergent American self-identity as a leader in technological advancement. And third, it seemed to promise a way to use the rapidly developing tool of aviation to full effect, but within the boundaries defined by ethical principles and deeply held convictions about American moral exceptionalism.

Despite the fact that a theory of precision bombing emerged early on, the ability to realize it lagged behind. For a variety of reasons, finding and hitting specific targets proved much more difficult than either theorists or practitioners had expected. This meant that for most of the twentieth century, plans and possibilities did not align. During World War II in particular, a considerable gap developed between what was expected of aerial bombing and what it could actually accomplish. This forced planners to engage in constant, real-time modifications of practice in order to extract as much from the long-range bombing tool as possible. The process was challenging; it pulled the Americans away from their theories, and away from their desire to use bombing within the boundaries and constraints they had articulated prior to the war. The post-1945 period saw a continuation of the quest for precision bombing; it resulted in incremental improvements that had, by the early twenty-first century, accumulated to considerable—and indeed remarkable—progress.

This essay traces the evolution of American ideas about precision bombing. It outlines the emergence of those ideas, and specifies the ways in which they were often out of step with technical possibilities. It explains the adjustments made to facilitate the use of long-range bombers in wartime, despite the challenge—throughout the twentieth century—of imperfect means of navigation and bomb-aiming. And it describes the ongoing and ultimately successful efforts to align the capabilities of bombers with preferred theories.

The real origins of American precision bombing theory are often overlooked. We are far too ready to dismiss World War I as a fallow era in the history of military ideas and military innovation. But this gets the story exactly wrong. Rarely was there a more active time in the development of military thought. And surely this was the case with respect to air war. There was not a single form of military aviation that did not receive at least a rudimentary trial between 1914 and 1918. If the Americans arrived late on the scene, they nonetheless paid close attention to the experience of their allies.

The British endured two rounds of German strategic bombing, first by Zeppelins and then by Gotha and Giant bombers. They waged their own reprisal campaign against the Germans in 1918, with the newly independent Royal Air Force (RAF). The British experience generated a great deal of thinking about the use of long-range bombers in war, and the Americans became particularly interested in the ideas of one British analyst named Lord Tiverton, a civilian attorney brought into wartime service. Tiverton, who possessed an analytical mind, argued that it would be most effective and efficient to seek out "bottleneck" targets in the German war economy—those on which the smooth operation of the enemy war machine depended most heavily. Tiverton's conceptualization called for selective targeting of particular locations; it was premised on aircraft being able to achieve a fundamental level of accuracy.

Tiverton wrote up plans, in September 1917, which articulated a comprehensive concept for long-range bombing.[1] These identified specific target sets, listing them by geographic regions. For Tiverton, it seemed self-evident that attacking the foundations of the enemy war economy would be the most efficient and effective means of using long-range bombers in war. In line with this, he prioritized such targets as the Benz motor works, the Bosch magneto factory and aeroplane works, petroleum refining stations, Saar Valley Steel works, chemical and munitions factories, and power stations.[2]

Major General Hugh Trenchard, former head of the Army's Royal Flying Corps, and, by the spring of 1918 the somewhat reluctant commander of the RAF's new "Independent Force" for long-range bombing, largely ignored Tiverton and the Air Staff. This was partly for political and idiosyncratic reasons that need not detain us here, and partly because he felt he simply did not have the types and numbers of aircraft necessary to implement the young Air Staff's grand design.[3] Instead, Trenchard mainly concentrated on tactical targets. But, because he was under public and political pressure to strike deep into Germany, he occasionally sent his aircraft to industrial targets and cities. Though the physical impact of the raids was modest, Trenchard nonetheless claimed that his bombing had important indirect effects on enemy war production, and that it had a psychological or "moral" (at the time pronounced "morale") effect on the enemy.[4]

While they fully accepted the idea of the "moral" effect of bombing, the Americans had a particular liking for the more analytical Tivertonian conceptions. Indeed, the program for strategic bombing they drew up but never implemented was based, nearly verbatim, on Tiverton's September 1917 plan. It was borrowed, two months later, by the American Air Service's Lieutenant Colonel Edgar S. Gorrell, who had been charged with designing a long-range bombing plan for the nascent U.S. air arm.[5]

As it turned out, the Americans entered the war too late to implement any plans for long-range bombing. Tensions within the U.S. Army over the role that aircraft ought to play in war also inhibited progress. Nonetheless, the Americans set up their own postwar bombing assessment so that they could glean the lessons the air war had taught. Like the British Air Staff, they criticized Trenchard for what they believed to be his rather haphazard and unsystematic approach to bombing.

The authors of the American bombing survey argued that in order to obtain useful results from bombing in future wars, it would be necessary to make a "careful study" of the different kinds of industries in an enemy nation, and ascertain "how one industry is dependent on another and what the most important factories of each are." The authors added: "A decision should be reached as to just what factories if destroyed would do the greatest damage to the enemy's military organization as a whole."[6] This conception established an early framework that would be reiterated and reinforced in the years between the two world wars. From the first, those Americans who thought hard about long-range bombardment gravitated toward the careful selection of key targets, and the accurate—or "precision"—bombing of such targets.[7]

After the war, after Versailles, and after its rejection of the League of Nations, the United States fell back into a defensive, isolationist posture. President Woodrow Wilson had asserted that Great War was the war to end all wars, and Americans citizens—who were much more concerned with domestic economic growth than with European politics—wanted badly to believe it. If General Billy Mitchell's demonstrations of air power captured the attention of the American people, they did not lead to the service independence that Mitchell so badly desired. Despite his best efforts, he was not able to convince his countrymen of the need for an independent air force; they simply did not feel threatened enough to overhaul and restructure their defenses, especially as they clung to the hope that war, in general, was a thing of the past. The American people did, however, grow increasingly "air minded" as they thrilled to the exploits of such aviators as Charles Lindbergh, Wiley Post, and Amelia Earhart. Commercial aviation burgeoned and by 1936 no less than half a million young men were members of the Junior Birdmen of America.[8]

In the meantime, theories of long-range bombing evolved slowly—but steadily— within the air arm of the U.S. Army. The U.S. Army Air Service, which in 1926 became the U.S. Army Air Corps, was held back from radical reform and had to settle for gradual and iterative change. Army leaders, still committed to the idea that aircraft should support land operations, were not inclined to give the aviators much scope for planning independent aerial operations. In addition, Secretary of War Newton Baker had, in his post–World War I report, attached moral opprobrium to the kind of long-range bombing that had been carried out in the Great War. He believed it to be inconsistent with American principles, and his own notions of lawful constraints on wartime behavior.[9]

All these things placed boundaries around what American aviators could say and write about the use of aircraft in future wars. But if they were constrained, they were not stifled. The Army's leadership did not manage to entirely contain the fertile minds of those who were, together, becoming a kind of insurgency movement within the ranks.[10] The lessons gleaned from the U.S. Bombing Survey were repeated in manuals, lectures, and training guides produced by the interwar Air Service/Air Corps: always, the primary emphasis was on identifying specific industries that were key to the enemy's war-making power.

Following in the steps of Tiverton, the U.S. Army Air Service/Air Corps produced its own theorists. William C. Sherman, who died young and is largely forgotten now, wrote a forward-looking book in 1926 titled *Air Warfare*. It outlined, in full, what would later become known as the "industrial fabric" theory of strategic bombing. Sherman asserted:

> Industry consists . . . of a complex system of interlocking factories, each of which makes only its allotted part of the whole. . . . Accordingly, in the majority of industries, it is necessary to destroy certain elements of the industry only, in order to cripple the whole. These elements may be called the key plants. These will be carefully determined, usually before the outbreak of war. . . . On the declaration of war, these key plants should be made the objective of a systematic bombardment, both by day and by night, until their destruction has been assured, or at least until they have been sufficiently crippled.[11]

By the end of the 1920s, American airmen were lured increasingly by the promise of the bomber and its role in the offensive theory of warfare, and they were ever more willing to run the risk of pushing the boundaries established by their parent service. When, in the summer of 1931 the Air Corps moved from Langley, Virginia to Maxwell Field, Alabama, it gained an added measure of distance from the Army—not only physical distance, but intellectual distance as well. Maxwell is where the real process of pulling away from the Army began. While the Maxwell theorists would pay lip service to the primacy of the ground war, their thinking moved in a different direction.

Instructors at Air Corps Tactical School (ACTS), men like Robert Olds, Haywood Hansell, Harold George, Kenneth Walker, and Donald Wilson began to refine and flesh out ideas about long-range independent bombing in war. Picking up in the tradition of Sherman, they imagined high-altitude bombing aimed at specific industries. The theory preceded the tools, but the tools were not far behind. In the 1930s the capabilities inherent in the new B-17 bomber and the Norden Mark XV bombsight both facilitated and reinforced the prevailing U.S. conception of air force doctrine: scientific and technological developments dovetailed not only with the existing emphasis on careful selection of targets, but also with the requirements of prevailing moral and ethical strictures. Ideas and machines came together, resulting in American plans for high-altitude, daylight, "precision bombing" of critical nodes in the enemy war economy.

Although the B-17 was authorized and designed for the defense of U.S. coastlines, its promise as a long-range bomber seemed self-evident to those willing to envision it in that role. Well-armed B-17 bombers—"Flying Fortresses"— would attack in formation and use interlocking fields of fire to defend one another. To those at ACTS, the industrial fabric theory seemed both intuitive and elegant. And the new tools would facilitate its implementation.

American airmen labored to derive scenarios and target sets, and to identify and assess the most vulnerable points of attack for an air force. This analytical outlook probably owed much to the progressive tradition and the strong appeal of Taylor's "scientific management" in the United States. Also, geopolitics allowed the Americans to take a more distanced, less emotional approach to the issue of air war than the Europeans, who could not escape the need to countenance future attacks on their cities. The American approach to bombing theory was reinforced by bitter lessons from the economic crash of 1929 and the subsequent Great Depression, which hit the United States particularly hard and seemed to reinforce the notion that complex, interlocking modern economies are readily subject to major dislocation.[12] Writing in *Colliers* magazine in 1939, Republican Congressman Bruce Barton gave voice to a widely held impression: "Industry is a living body, highly integrated, with nerves extending to every part. Cripple a limb, and the whole body limps, prick even a finger, and the pain is felt throughout."[13]

The American preoccupation with science and technology could not resolve the many unforeseen problems inherent in the planning process however. Theory was undermined by some flawed assumptions and by a failure to fully comprehend the very heavy reliance it necessarily placed on high-quality and time-sensitive strategic intelligence. Also, overarching faith in American technology seemed to crowd out any concern that very accurate bombing might prove difficult—even impossible in some circumstances. Committed to an idea, the airmen did not fully confront those things that might undermine it. In the 1930s, test results "showed that bombing was accurate only in excellent weather against a clearly outlined, undefended target in the middle of wide-open terrain." Even as war loomed on the horizon, bomber crews rarely dropped bombs from heights above 12,000 feet.[14]

But this was not the only potential problem facing American bomber advocates. Being able to attack selected targets and bomb them precisely rested on the assumption that one would be able to reach them without sustaining prohibitive losses in the process. As advocates of independent air power found their stride, they increasingly began to brush aside the candid observations about bomber vulnerability that had appeared in American writing just after World War I. Some of this was understandable since, by the late 1920s bomber speeds were well ahead of fighter speeds and seemed likely to stay that way. But as defensive techniques made strong advances in the late 1930s, bomber advocates could not bring themselves to fully admit that these called into question the foundation on which their ideas rested.

In the United States, Air Corps planners struggled with the idea of long-range fighter escorts. At the end of the day, though, they could not see how to design a plane that could keep up with the bombers over long distances, and then compete on equal terms with enemy short-range fighters. The problem posed inherent and daunting technical challenges. Various pieces of the puzzle were out there in one form or another—such as auxiliary tanks—but they were never adequately assembled in the minds of planners prior to the war. There are lots of reasons why this is so, although the extant literature (my own book included) has not yet offered us a completely satisfying answer as to why neither the British nor the Americans came fully to grips with the questions of bomber penetration prior to the outbreak of World War II. A major part of the problem on the American side, at least, was that planners did not put the issue right at the top of their priority list because they did not feel they had to: so long as theorists could assume that bombers could self-defend, the escort issue did not make or break the theory of high-altitude, daylight precision bombing.

Not even the grim experience of the British early in the war deterred the Americans from the idea that precision strikes by high-altitude bombers flying in formation would be the path to success. The British, the Americans concluded, had taken the wrong approach to the task, and had not shown the fortitude necessary to bring daylight bombing to its fullest potential. From a comfortable distance, American planners held fast to their assumptions and criticized their allies for a wrongheaded diversion into inefficient nighttime "area bombing."[15] But the British too had wrestled with the long-range escort problem, and had come up short. When, in 1941 Chief of Air Staff Sir Charles Portal told Prime Minister Winston Churchill that a long-range escort was not in the offing, the prime minister was distraught. He seemed to understand, perhaps more clearly than anyone else did at that time, just what the failure meant: "this closes many doors of hope and opportunity," he remarked morosely.[16]

Facing heavy losses in their attempts to penetrate into Germany in daylight, the British switched to nighttime attacks by heavy bombers. Bombers flew alone in narrow streams designed to pierce through, and evade German defenses. Convinced that the Americans would ultimately be forced into the same unhappy decision, the British—Churchill in particular—tried to persuade the Americans to simply join the night offensive from the outset. Refusal by the Americans signaled the faith they had in precision daylight bombing, and in the notion that groups of heavily armed bombers could defend one another. The prime minister understood that the problem of penetrating German defenses was rather more complex than the Americans were allowing themselves to believe; he was deeply concerned about the American choice and, in his private correspondence with his air chiefs, foresaw most of the problems the Americans would face in 1943.[17]

An abundance of American optimism and "can-do spirit" hyped the promise of American daylight bombing as the U.S. bombing effort was just beginning to get off the ground. Yankee criticism of British area bombing—at a time when Bomber Command was routinely taking hundreds of bombers to targets while the Americans were barely taking dozens—must have been grating to the British to say the least. And the high-minded assertions of moral superiority with which the Americans often imbued their comments must have been anything but endearing to their battle-scarred allies. But the Americans were committed to their own ideas, and were stubbornly determined to try them out. In hindsight, it is probably a good thing that the Americans were as single-minded and determined as they were—for reasons that will be explained below.

Through most of 1942, the Americans flew only limited raids over occupied territory. Since the full challenge of flying deep into Germany lay ahead, they had encountered little—by the end of the year—to cause them to overturn their initial assumptions. At the Casablanca Conference in early 1943, Allied planners accommodated the Anglo-American difference of views, agreeing to disagree on a bombing directive that let the two nations go their separate ways: British nighttime area bombing would continue, while American daylight "precision" bombing would commence over Germany. The slogan "round the clock bombing" nicely papered over the cracks. Ultimately, as the British official historians have pointed out, the 1943 Allied air offensive was more a competition than a "combined" bomber offensive.

United States Army Air Forces (USAAF) bombers began flying over Germany proper in early 1943; the year would prove to be a fateful one as the Americans came face to face with the very problems their British allies had warned them about. Both target acquisition and bomber penetration proved harder than the Americans had expected, upsetting the assumptions on which the theory of daylight precision bombing had rested. In mid-April 1943, 16 of 115 bombers dispatched were destroyed in a raid aimed at the Focke-Wulf plant in Bremen; a further 44 were damaged. In response to American efforts to wrest control of their air space, the Germans set up a defense that, by the summer of 1943, contained a fighter command system covering a geographical zone up to 480 miles deep.[18]

The Americans faced production shortfalls for their bombers, while the Germans rapidly expanded their fighter force. And American attempts to find and strike specific targets were undermined by the pervasive cloud cover over Northern Europe, and the altitudes and evasion techniques forced upon the bomber pilots by German FLAK (*Fliegerabwehrkanonen*)—fired by increasing numbers of anti-aircraft guns. In addition, the much-vaunted Norden bombsight, which had given good results in controlled tests back in the United States, had distinct operational limits in overcast Europe. The Eighth Air Force began to divert with increasing frequency to area targets or fringe targets less likely to extract a high toll in aircraft and crews.[19]

In late 1942, General Henry "Hap" Arnold, commander of the Army Air Forces, had requested investigations into navigational aids used in homing, and target location by radio means; early in 1943, individual bomb-sighting was abandoned and replaced by a procedure in which all aircraft dropped their bombs simultaneously on a signal from the leader of a combat "box" of 18–21 aircraft.[20] The goal was for aircraft to bomb in tight patterns; but enemy fighters, anti-aircraft fire, and smoke frequently thwarted the desired result. By late 1943, General Arnold permitted American bombers to operate "blind"—to attack area targets through cloud. While he knew that this would greatly increase the operating tempo of U.S. forces, he tried to distance himself from the uncomfortable language, emphasizing that a term like "bombing through overcast" be used instead of "blind bombing."[21]

As we look back on the methods of navigation and target acquisition available in the early to mid-1940s, it is sometimes hard to realize just how primitive they were compared to what is available in the twenty-first century. All the methods put heavy demands on navigators and bombardiers—both of whom, like the other members of the crews of World War II bombers, had to operate in gelid, noisy airplanes that were not infrequently in the crosshairs of enemy defenses. Radar was used to aid both navigation and target acquisition, but at no time during World War II was a complete substitute found for dead-reckoning. This simple statement is nonetheless terribly important because at the heart of dead-reckoning is the ability to do frequent and accurate calculations of the ever-changing wind vector—and calculating the wind vector was an inherently difficult problem for air crews.

Prior to 1942, when Bomber Command first tried to hurt Germany through bombardment, there was no accurate and reliable method of dead-reckoning. Citing this fact, the British official historians have written: "What is surprising about the years before 1942 is not that so many crews failed to find their targets, but that more of them did not fail to find England on their return."[22] Radar aids developed and, while they were far better than nothing at all, each one had its own set of drawbacks and constraints. Forced to rely on them early, British crews grew increasingly adept at using the technological means available to them.[23]

In the meantime, other elements of American operational practice proved untenable. An August 1943 raid against ball-bearing works in Schweinfurt, Germany, and the Messerschmitt factory in Regensburg saw 16 percent of the attacking force lost. Another raid against Schweinfurt in October cost 198 planes shot down or damaged out of a total force of 291. In four raids carried out over six days in October, 148 American bombers failed to return to their bases.[24] A reckoning was now unavoidable. Since it was too costly to fly unescorted bombers into Germany by day, the Americans would have to change tactics—or change targets (as the British had before them). In the event, they chose the former. Relying on a large and expanding American production base, they built long-range fighter escorts with auxiliary fuel tanks (that could be jettisoned after use), and sent them to Europe in large numbers. These escorts would range over enemy territory, providing the protection that bombers could not provide for themselves.

Flying to targets the Germans felt compelled to defend, American bombers drew *Luftwaffe* fighters into the skies, forcing them into contests with the long-range escorts. In vast battles of attrition over the skies of Europe, the Americans eventually won dominance over the German Air Force. But the outcome was achieved after a struggle to backfill around a prewar theory that had gaps and problems. Interwar planners had not anticipated using bombers as bait. But war was a harsh teacher, and the Americans found themselves with no choice but to learn in real-time, and change direction in mid-course. In this, the new head of the United States Strategic Air Forces, General Carl A. Spaatz, revealed flexibility and adaptability as he developed successful methods of inter-operating bombers and fighters.

At the end of the day, the relentless pressure placed on the *Luftwaffe* by bombers and escorts greatly aided the British, who also were struggling by late 1943 due to the increasingly lethality of German night fighters. Had the Americans not resisted earlier British pleas to join the nighttime area offensive, it is not clear that the Combined

Bomber Offensive would have found a way forward in 1944. Though American resistance to British pressure had been instinctive, in this particular instance it paid vital dividends.

The gradual erosion of the *Luftwaffe* surely made it easier for American bombers to reach targets unmolested. But long-range escorts could not solve the problem of European weather, which remained the primary constraint on the performance of American bombers after the *Luftwaffe* had been largely defeated. Pressure to bomb through cloud forced the Americans to rely on their few bombers equipped with the H2X (derived from Britain's H2S) radar navigation and blind-bombing aid. But the H2X program faced problems, especially at the outset. Equipment failures were common, due to inadequate maintenance of the highly technical equipment—and these were exacerbated by a lack of parts and maintenance personnel.[25]

Forced to operate at night early in the war, Bomber Command crews had grown steadily proficient in instrument flying and the use of a variety of navigation and target acquisition aids.[26] Thus, in poor weather conditions (which prevailed much of the time over Europe), British crews outperformed American crews. By the spring of 1944, when both British and American strategic bombers were pulled under the command of General Dwight D. Eisenhower in preparation for the Normandy landing, British bomber crews were able to hit a wide range of specific targets. American crews also improved their technique over time, and their success was aided, naturally enough, by the waning of the *Luftwaffe*. But the Americans continued to be thwarted by bad weather conditions. In non-visual bombing from October to December 1943, the Americans recorded accuracy rates no better than those achieved by the British in the summer of 1941, when the grim statistics of the investigation known as the "Butt Report" had revealed that Bomber Command typically got only about one in five of its aircraft within five miles of the target. A March 1945 Anglo-American conference on accuracy revealed that in conditions of heavy cloud cover (which included the majority of missions in the winter months, 1944–45), Eighth Air Force bombers still placed 42 percent of bombs more than five miles from the target. And, of those inside the five-mile radius, the average circular error was 2.48 miles.[27]

In heavy cloud cover, the Americans tended to attack marshaling yards—large identifiable targets in cities. These raids, which often contained significant numbers of incendiary bombs to maximize collateral damage, tended to be large and destructive. In the Far Eastern theater, clouds and the winds of the jet stream played havoc with American attempts to bomb factories and industry from high altitude; by February 1945, the new field commander of the 20th Air Force, Major General Curtis LeMay, was moving toward nighttime, incendiary attacks on whole Japanese cities.[28]

Aside from statements of intent, there was often little to distinguish between the practical effects of British area bombing and American "precision" bombing. Still, the Americans adhered to the language they had used going into the war. This was in part because they sought to define themselves in contradistinction to the British, and in part because this conceptualization of precise, efficient bombing really was—and had remained—part of the fabric of Air Service/Air Corps/USAAF identity. Whenever weather permitted, the Americans returned to their preferred choice of specific targets in the enemy war machine.

The selection of industrial and military targets was done carefully, and was based on the ongoing efforts of targeting staffs operating on the basis of the best available intelligence. Perhaps even more so than other military operations, aerial bombing is tremendously sensitive to accurate and reliable intelligence. But what was available to planners during World War II was not always comprehensive or timely, and errors resulted from these shortcomings. In Britain, the early efforts of the Ministry of Economic Warfare to find the real Achilles' heel in the German war economy were eventually dismissed by the commander-in-chief of Bomber Command, Sir Arthur Harris, as a fruitless search for "panaceas." Indeed, his loss of faith in this process caused him to call into question nearly all intelligence information offered to him (by those other than his own staff) late in the war.

The Americans operated on the basis of some flawed assumptions. Like the British, they assumed that the German war economy was at full tempo early in the war when, in fact, it still had plenty of excess capacity. They assumed too that the Germans would not be able to readily substitute for scarce materials, stockpile critical materials, or disperse industry. In each of these realms, however, the Germans proved masterful; the creative leadership of Albert Speer, head of the Ministry of Supply, enabled the Germans to repeatedly disperse and reorganize their assets and resources. In addition, the Allies missed some opportunities they might have taken advantage of more fully. Specifically, the electrical power net and the powder and explosives industry were more vulnerable than the American planners assumed.[29]

By 1944, General Spaatz had concluded that synthetic oil was a genuine vulnerability for the Germans. His instinct in this regard was backed up by sound intelligence indicating that the Germans frequently expressed anxiety about their sources of this essential commodity. Eisenhower was sympathetic to Spaatz's outlook, and gave him permission to strike oil targets whenever his bombers were not strictly required for duties related directly to D-day and the break-out from the Normandy beachhead. By the summer and early fall of 1944, Anglo-American attacks on German synthetic oil production facilities—in conjunction with the loss of eastern European oil sources due to the advance of the Red Army—had seriously undermined Germany's ability to wage maneuver warfare. Only the onset of very bad weather in the autumn, forcing a pause in the intensity of oil attacks, enabled the Wehrmacht to recover enough to launch a counteroffensive in December, the Battle of the Bulge.

But even as they closed in on German oil, the Americans made only incremental progress. The United States Strategic Bombing Survey (USSBS) would later reveal that, in "precision" strikes against German oil facilities, the American strategic air forces would place only 2.2 percent of total bombs dropped on crucial buildings and equipment; over 87 percent of bombs were "spread over the surrounding countryside," outside the perimeter of the plants. Repeat raids were a necessity, and slow progress was all that could be expected under the circumstances. Bomber Command crews contributed in important ways to the oil campaign, but their efforts too were hindered by the adverse weather that beset northern Europe in the autumn.[30]

At the end of the day, oil raids and the steady Anglo-American hammering of the German transport net proved the most effective uses of long-range bombers. Both targets offered synergistic effects with the ground war: the Red Army's westward advance cut off alternate oil supplies in the east, and attacks on both oil and railways gradually strangled the Wehrmacht's ability to wage a war of maneuver. Tactical aircraft, which had been moved on to the continent after the Normandy landing, also contributed to the air war. The requirement to interdict supplies close to the battlefield, and to support troops on the ground, naturally put a premium on accuracy. When heavy bombers were employed in these roles, as they sometimes were, they revealed the blunt, bludgeon-like quality of their 1940s-era technology. On 25 July 1944, heavy and medium bombers flying in support of Operation Cobra bombed short, killing 102 friendly soldiers (including Lieutenant General Lesley J. McNair), and wounding a further 380. The casualties mainly affected the American 30th Infantry Division.[31]

In 1945, the Americans bombed a wide range of targets. Some of these attacks were designed to further erode the German war economy, some to aid the advance of the Red Army, and some to hasten the collapse of German will. The latter two categories typically involved strikes on cities and towns rather than industrial facilities specifically. This move away from preferred doctrine signaled the urgency the Americans felt about bringing the war in Europe to a conclusion before the war in the Far East reached a crucial phase, including the invasion of Japan's home islands.

On 24 November 1944, the 21st Bomber Command in the Far East had carried out its first large-scale attack on the Japanese homeland. Under General Haywood Hansell, the 21st struggled against the heavy cloud cover and high-altitude winds prevailing over Japan. Hansell, a product of the Air Corps Tactical School, had a tenacious commitment to selective industrial targeting, but his efforts to follow his preferred course were repeatedly thwarted. In January 1945, a frustrated General Arnold replaced him with General Curtis E. LeMay. A skilled field commander with a reputation for solving tough problems, LeMay was well aware that Arnold wanted him to find a way to make bombing work in the Far East. The weakness of Japanese air defenses allowed LeMay to try a new tactic: he stripped his large B-29 bombers of their armament, loaded them with incendiary bombs, and flew them in low and at night against vulnerable Japanese cities. Before atomic weapons were dropped on Hiroshima, LeMay's crews had waged 67 such raids. One such fire raid in Tokyo, on the night of 9–10 March 1945, killed 80,000 city dwellers in one night.[32]

In both the European and Far Eastern theaters, the barriers to precision bombing had pushed the realities of American long-range bombing well away from interwar theory. In practice at least, British and American bombing converged a good deal of the time. But in Europe, the Americans never sought to incite the large firestorms that Bomber Command tried to achieve each time it hit a city. And in both the European and Far Eastern theaters, the Americans always returned to their default targets—key resources in the German war economy—whenever weather permitted. Unlike Sir Arthur Harris, the Americans did not believe that cities were the most profitable enemy targets to strike. But the phrase "precision bombing," which the Americans used rather too liberally in an effort to distinguish themselves from the British, requires a great deal of qualification when it is applied in the context of World War II. Only in a few extraordinary instances did either the Americans or the British achieve what could be considered "precision" bombardment (according to a dictionary definition of that word).

American postwar surveys and evaluations, including the USSBS, tended to emphasize the utility of aerial attacks against oil in the European theater. This emphasis reinforced the idea that prewar theories of selective attack against key industries were sound, and had been vindicated by the experience of the war. The USSBS was stingy in offering credit to Bomber Command; it criticized area bombing and gave little attention to the important contributions the British had made to the oil campaign and to the pre-Overlord bombing of specific targets in France. Characterization of the USAAF as the "precision" air force, in contrast to the British, allowed the Americans to maintain and further conceptions of their own self-identity, and more easily deflect postwar arguments about the indiscriminate nature of long-range bombing.[33]

The data, analyses, and potential lessons contained in the USSBS were not used to their full advantage of the aftermath of the war. This was partly due to the structure of the Survey itself. Containing over three hundred separate reports, it was unwieldy and sometimes inconsistent. The summary reports, which were the ones most widely read and quoted (oftentimes the *only* ones read and quoted), had the ring of least common denominator conclusions attempting to generalize across too broad a range of views and issues. Debates over aerial bombing in World War II—which commenced soon after the war and have never abated—were inevitably drawn into the politics of the Army Air Forces' postwar quest for independence. This fact pushed the not-yet-separate Air Force away from a searchingly rigorous investigation of its own record since the cost of such an honest self-examination might have been unacceptable with respect to the future of the service, and its claim on roles and missions. Finally, the advent of atomic weapons made close scrutiny of the past seem rather an idle task. In the midst of an apparent revolution in military technology, much of the Bombing Survey seemed obsolete even as it rolled off government printing presses in Washington.

General Arnold was anxious to claim that the atomic bombings had ended the war in the Far East; he was equally anxious to embrace the USSBS summary conclusion that, "Allied air power was decisive in the war in Western Europe." Both statements would not only help to secure the postwar independence of the USAAF, but would help to set up expectations for the use of air power in future wars. But "decisive" is an elusive term. What was surely apparent was that no modern military could afford to do without a sophisticated and capable air force. Once the Allies had wrested control of the air from the Germans and the Japanese, victory was virtually assured for them;

they had only to face and endure the last, fierce blows of a mortally wounded enemy. Equally apparent was the fact that accurate and sustained bombing of key targets could be, ultimately, devastating to one's enemy.

Identifying those targets with certainty, however—and then striking them accurately and in a sustained way—was far harder in wartime than the air power advocates had anticipated. Both war economies and societies had proven to be more robust and resourceful than they were expected to be. Even in the midst of heavy bombing, the German war economy in particular managed to achieve substitution of goods and commodities, re-allocation of resources, and physical movement of assets. Societies proved too that they can absorb tremendous punishment under certain conditions. In Germany, this was due in part to the robustness of coercive measures to restrain dissent, and fear of the consequences of defeat (intensified by heavy propaganda); in Japan it was due to an overwhelming national commitment to the emperor, bolstered by the religious influence of Shintoism and the cultural influence of the Bushido code.

Postwar: Atomic Weapons and Conventional Wars

During the summer of 1945, special crews of the 20th Air Force in the Far East trained for months for a secret mission. Its details were so tightly held that when, in early August, the *Enola Gay* lifted off the runway at Tinian Island, its personnel knew only that they were on an unusual sortie with unique demands. Little about the postwar future was entirely clear to the American policymakers who opted to drop two revolutionary weapons on two Japanese cities. Seeking a prompt end to a tenacious fight that threatened to drag on into a disastrous final bloodletting, President Harry S. Truman and his advisors oversaw the culmination of a program that had begun many years before, with a different rationale.

The atomic bombs dropped on Hiroshima and Nagasaki were delivered by air. Atomic weapons would, for the immediate future, remain bomber-delivered weapons; and they would, for the next several years, remain uniquely American weapons. During those years, the Americans and the Soviets would discover seemingly irreconcilable political differences, and the resulting "cold war" would shape world politics for nearly half a century. It would, as well, shape the future of what would become the United States Air Force (USAF).

In the late summer of 1945, however, nothing about the Cold War seemed foreordained. With the end of fighting the Americans promptly demobilized and made radical cutbacks in defense expenditure. But isolation behind the Atlantic and Pacific moats was no longer an option for the Americans. The need for a permanent military infrastructure was apparent, and the first order of business was to revisit the question of defense organization, and the assignment of roles and missions to the individual services. An independent United States Air Force was created as part of the National Security Act of 1947; the new service's chief officer would have co-equal status with the heads of the Army and Navy, under a new bureaucratic structure that created a civilian secretary of defense at the helm of what was at first called the "National Military Establishment," and later the "Department of Defense."[34]

The postwar argument for a separate air service in the United States rested—as it had in Britain many years earlier—on the assertion that future wars would involve aerial bombing, and that an autonomous air force would be best equipped to plan and implement aerial attack and defense. The structure of this argument necessarily placed a heavy emphasis on strategic bombing as a mission of signal importance to the security of the United States, and would set the USAF on a trajectory that would highlight that mission over all others.

As the sole possessor of atomic weapons, the U.S. Air Force may have seemed a formidable force on paper. But funding cutbacks, demobilization and restructuring, and a focus on the fight for independence (rather than training and operational readiness) meant that the real strength and capability of the USAF was strictly limited in its early years. Even the high-profile arm of the service, Strategic Air Command (SAC), struggled until it was rejuvenated in 1948 by the hard-driving and demanding General LeMay.

In the immediate postwar years, targeting for strategic bombing was something of an ad hoc affair. Special arrangements were made to modify British airfields so that they could, in an emergency, accommodate American B-29 bombers and their nuclear munitions. Through 1948, the number of available nuclear munitions was small. President Truman was so ambivalent about the prospect of nuclear weapons that in response to the first atomic target lists, he asked planners to produce an alternative, conventional plan.[35] Early postwar Air Force targeting reflected the difficulties of planning and executing air strikes against the Soviets in light of a scarcity of both weapons and information. Aim points were cities ("urban industrial concentrations") since these were the only targets that promised any hope of successful identification.[36]

LeMay and his Air Force colleagues moved forward nonetheless, basing their assumptions and their methods of training on their experiences during World War II. Breakthroughs in technology allowed the number of available nuclear weapons to expand more rapidly after 1948 and, by 1952, LeMay's crews were training—with radar reconnaissance methods, ultrasonic maps, and new navigational equipment— for "blunting" counterforce missions against Soviet airfields, even though the precise coordinates of those fields were not known. After 1949, when it was known that the Soviets had acquired the ability to produce atomic weapons, it became essential for the Air Force to develop any means available for finding and hitting their delivery systems, so as to eliminate (or, later, reduce) the amount of damage they might inflict on the United States. LeMay developed methods for the aerial reconnoitering of Soviet borders, eventually over-flying Soviet airspace in search of strategic targets. Between 1949 and 1953, however, most of the Air Force's intelligence on Soviet targets was acquired through Project WRINGER, which used civilians in Germany, Austria, and Japan to interrogate thousands of repatriated prisoners of war who were returning from the Soviet Union.[37]

In the meantime, the U.S. military was caught rather flat-footed by a need to respond to an urgent crisis on the Korean peninsula. Largely unprepared to fight a conventional war, the Air Force did what it could to stem the North Korean assault across the 38th parallel and to aid in the restoration of the status quo antebellum. LeMay believed that the North Korean aggression might simply be a sideshow—a diversion from a potential Soviet aggression in Europe. He did not want his SAC bombers "diverted" to Korea, and he wanted those bombers in the theater to largely repeat his World War II campaign in the Far East by waging fire raids against North Korean cities. The commander of the United Nations forces, General Douglas MacArthur, denied his request at first. But when the Communist Chinese entered the war in November, Air Force heavy bombers were unleashed on a wide range of targets, including North Korean cities.[38] Still, the Air Force faced a conundrum in that most of the genuinely strategic targets—the sources of supply—were located in China and the Soviet Union, and these remained off-limits to the bombers of the Far Eastern Air Force (FEAF). After the war ended, LeMay would remark, "We never did hit a strategic target."[39]

Soviet-built MIG fighters complicated the issue of bomber vulnerability, and it was not long before American B-29s were forced to fly mainly at night. This led to a good deal of consternation at SAC headquarters, and LeMay—an expert problem-solver—set about making modifications that would help his B-29 gunners. But, because he was not prepared to reveal, in Korea, technologies such as chaff and certain jamming techniques that might prove crucial in an "all out war," he did not protect his bombers as much as he might have done.[40]

Throughout most of the war, FEAF engaged continuously in operations—such as interdiction and close support—that had not been high on the postwar Air Force's priority list. The Air Force's approach to (and philosophy of) close air support was rather different than that of the other services—a fact which caused consternation and interservice strife during the Korean War, and has done so ever since.[41] Still, these activities put a premium on precision, driving the Air Force to enhance its technology and improve its techniques in that realm.

Razon bombs, which had their provenance in World War II Azon bombs, could be guided by radio signals that would vary the pitch of their fins to reduce deflection errors. Though their use got off to a slow and halting start in Korea, their reliability improved over time, and they were used with good success to drop bridges. These were the predecessors of the precision munitions the USAF would use in later twentieth-century conflicts.[42] Short-range navigation radar (SHORAN) capability was developed, in 1950, in anticipation of the possible use of atomic weapons for the support of troops in Korea. Because it allowed crews to determine and know their navigational position with a good deal of accuracy, it proved helpful in Korea (especially since normal radar scope bombing was affected by the lack of terrain contrast). But though SHORAN was the most accurate method for B-29 non-visual bombing, it had drawbacks (like its World War II predecessors) that made it difficult to use, and impractical in certain circumstances.[43]

In general, the precision achieved in the Korean War was an improvement over that achieved in World War II. But the difference was not dramatic. As Conrad Crane has pointed out, "bombs dropped from a four-ship diamond formation thoroughly covered a rectangle about 500 feet wide and 2,000 feet long." Errors, he explained, were made all the more likely by poor maps and target photos, fatigued crews (that, early on, had to load their own ordnance), and left-over, World War II bombs that had been in storage long enough to upset their normal center of gravity.[44]

Because it was heavy, however, even inaccurate bombing imposed fearsome damage and terrible suffering on the North Korean people. During the course of the war, eighteen of twenty-two major urban areas suffered at least 50 percent obliteration. Witnesses claimed that the towns had been turned to rubble, and that most all North Korean families suffered dead or wounded as a result of the bombing.[45] As negotiations toward a peace settlement dragged on in 1952, FEAF was given a broader mandate to attack a wide array of targets in an "air pressure" campaign to bomb military targets that would, by their location, have a deleterious impact on "the morale of the civilian population actively engaged in the logistic support of enemy forces." The final phase of the campaign was particularly dramatic: three dams situated near railway lines were struck hard; the raids produced flooding in nearby villages and rice fields. It is not clear whether these raids hastened the conclusion of the talks; the evidence is ambiguous. But the campaign in Korea offers another example of an air force widening its target list when earlier bombing efforts are either frustrated or do not produce the results sought.[46]

After the Korean War ended, the Air Force moved quickly back to a focus on what it viewed as its primary task: deterring and, if necessary, fighting the Soviet Union. FEAF's 1954 final report on the war concluded that the Korean conflict contained so many unusual factors as to make it a poor model for planning. It argued, as well, that the "lavish close air support" provided in Korea ought not necessarily to be expected in future conflicts.[47] Deterring or (if necessary) fighting the Soviet Union was the USAF's conception of its *raison d'être,* and the service was more than happy to return to that familiar ground. The Korean War had loosened up the purse strings for postwar military spending, and the Air Force benefited accordingly—especially Strategic Air Command. General LeMay became vice chief of staff in 1957, and chief of staff in 1961.[48] By 1964, three-quarters of the Air Staff's upper echelon came from SAC. The American nuclear arsenal continued to grow (from 1,750 weapons in 1954 to 26,500 by 1962), and SAC prepared to deliver them in a "massive, pre-emptive bomber assault."[49]

Under President Dwight Eisenhower's "New Look" policy, defense economies were achieved by placing the main burden for U.S. national defense upon the nation's nuclear arsenal—and thus SAC. General LeMay had a relatively free hand in planning for fighting a nuclear war, and his plans were oriented to speed and mass.[50] SAC's basic war plan of March 1954 called for 735 bombers to hit Soviet early warning screens from all directions simultaneously in order to defeat Soviet defenses. LeMay wanted as little left to chance as possible, and he wanted—in line with standard air force doctrine—to "blunt" as quickly as possible, the Soviet capacity to fight back. SAC's tendency towards large and overwhelming war plans drew criticism at times from those who felt that the president would, in a crisis or wartime scenario, need more flexible options. In mid-1954, Eisenhower himself urged planners to envision ways in which the United States might fight the next war in such a way as to "attain our national objectives with the minimum cost and the least dislocation to the world."[51]

LeMay's desire to blunt Soviet warfighting capacity and to knock out Soviet defenses pushed him constantly toward seeking out means of effecting bombing accuracy, even in the context of nuclear warfare. By the end of the 1950s, technological advancements and improved intelligence brought counterforce attacks increasingly into the realm of the possible. The development of intercontinental ballistic missiles (ICBMs) pushed in the same direction. Increasing Soviet investment in ICBMs put a premium on SAC's ability to identify and attack (with bombers and its own ICBMs) Soviet missile fields and supporting infrastructure. In 1960, a newly organized Joint Strategic Target Planning Staff went to work on what would become, in 1961, the first Single Integrated Operational Plan for nuclear war; it was designed to serve either as a response to Soviet attack, or as a preemptive strike. Though the great majority of its listed targets were classified as "military targets," it was still a massive plan that would have resulted in millions of Soviet casualties.[52]

President John F. Kennedy and his secretary of defense, Robert McNamara, were not comfortable with the LeMay-influenced version of strategic defense. They wanted more flexible options, including less reliance on the threat of wholesale nuclear war, and a more fully developed set of conventional options. In addition, McNamara rejected the idea that the United States would ever initiate a nuclear exchange. He argued instead that the robust nuclear capability of the nation—because it could survive a Soviet first strike and deliver an overwhelming retaliatory blow—would be used instead to deter the Soviets by making it clear that any initiation of nuclear war would insure grievous harm to themselves. Since it would cost less to maintain a survivable retaliatory force than to build a nuclear force designed to provide, as General LeMay put it, "decisive counterforce potential," McNamara used this concept of "Assured Destruction" as a lever for bringing the size and cost of U.S. expanding nuclear forces under control.[53] But the theory and policy of "Assured Destruction" drew some resistance from Air Force planners who, unsurprisingly, preferred to provide deterrence through detailed and comprehensive war plans that accounted for the full range of the enemy's target set. Even though Kennedy's "Flexible Response" strategy enhanced American conventional forces and gave the president more options for the use of force, U.S. nuclear war plans generally remained large and overwhelming, in the tradition of LeMay and SAC.

Despite his strong interest in reforming U.S. nuclear strategy, Secretary McNamara's time was increasingly devoted to the deepening crisis in Vietnam, and to the early stages of what would prove to be a lengthy American war with that divided nation. Once again, a limited, conventional conflict in Asia was not what Air Force planners had expected. Indeed, despite ongoing conflict in Asia through the 1950s, the *Air University Quarterly Review* published only two articles (in the whole of the decade) relating to air power and insurgency movements there.[54] And, once again, the Air Force would find itself frustrated by the nature of the conflict, and by the constraints and boundaries inherent in it. An ideological battle that pitted North Vietnamese Communists and South Vietnamese insurgents against South Vietnam's indigenous army and the American forces supporting that army, the war was, until its late stages, unconventional in nature—even though the Americans tried hard to fight it as a conventional war. The enemy did not rely on sophisticated industrial power, or on an advanced, mechanized army to achieve its aims. In addition, key enemy sources of supply came from the Soviet Union and China, both of which were off-limits to U.S. target planners.[55]

Turning to air power for what they hoped would be a swift and efficient resolution of the increasing insurgency movement, the members of the Johnson administration leaned toward a program of "graduated and continuing reprisal" in response to enemy aggression. Beginning in late February 1965, the administration commenced what would be a four-year bombing program of aerial interdiction, Operation Rolling Thunder, characterized by increasing pressure against North Vietnam. But the North Vietnamese managed to deflect much of the impact of Rolling Thunder by dispersing industry and supplies, and by developing a powerful air defense system. The gradual implementation of Rolling Thunder allowed the enemy forces to learn from their own mistakes: they had time to make adjustments that would enable them to better resist the impact of the bombing. Pressure by the Joint Chiefs to escalate the timing of the air campaign, and to broaden the target base generally, met with resistance from President Lyndon Johnson, who wished to keep the war off the front pages of newspapers, and to keep the American people focused on domestic issues, including his "Great Society" project.[56]

Both at the time and after the war ended, many members of the Air Force argued that the war against North Vietnam and the South Vietnamese insurgents could have been won largely by air power if the Air Force had been able to fight the war the way it wished to—using overwhelming force to break the will of the enemy early on. This debate is unlikely, ever, to be fully resolved. If we know that the gradual, iterative warfighting strategy chosen was the wrong one, it is not clear that an alternative strategy would have been more successful. By late 1967, most of the targets on the JCS target list had been destroyed; by the end of the war, the U.S. Air Force had dropped some 6,162,000 tons of bombs on Vietnam—vastly more than had been dropped by the Allies in all of World War II. Just as the British found out during the American Revolution, an insurgency movement presents few targets that are genuinely strategic—that are so central to the war effort as to decisively affect its outcome. In addition, insurgents can control the pace of conflict; when the pressure is raised, they can simply merge back into the social fabric and wait for an opportune moment to resume their struggle.[57]

The frequency of interdiction and ground support bombing required of the Air Force in Vietnam once again placed a premium on accuracy. The Air Force sought, throughout the war in Vietnam, to improve its capability in this realm. By 1972, the Air Force was in a position to use new precision-guided munitions, including electro-optically guided bombs and laser-guided bombs. This development would profoundly affect the use of air power throughout the rest of the twentieth century, and into the twenty-first century. These new munitions greatly increased accuracy and, during the Linebacker I campaign (from May to October 1972), they were used to attack bridges and other pinpoint targets. The Air Force also implemented an effective long-range navigation system for North Vietnam. Like the SHORAN used in Korea, it enabled aircraft to determine their position with a high degree of accuracy. It was not, however, as effective as its Korean War predecessor because the distances from its transmitters in South Vietnam were longer than the distances covered by SHORAN in Korea.[58]

Fighting an able and increasingly sophisticated opponent in the skies over Vietnam also pushed the Air Force to constantly refine its tactics and technology for penetration and survivability. Sophisticated jamming equipment helped neutralize the effect of improved North Vietnamese radar; and agile evasion tactics helped Air Force and Navy pilots evade SA-2 surface-to-air missiles. Developments in electronic warfare, including the use of EC-121 aircraft and Navy ships, enabled U.S. forces to track Russian-supplied MiG fighters and alert pilots to their presence. Command and control, routing, and escort and support tactics evolved constantly.

During the Linebacker II campaign of December 1972, B-52 strikes—which were planned and scheduled by SAC Headquarters in Omaha, Nebraska—revealed the high degree of technological sophistication inherent in the bomber tactics of the day. This campaign, which saw formations of aircraft occupying more than 70 miles of airspace (including heavy bombers and the large numbers of support aircraft necessary to help them penetrate enemy airspace), would nonetheless mark the end of massed heavy bomber formations for strategic bombing. Afterwards, no air force could afford to build and maintain such large numbers of heavy, multi-engined aircraft.[59]

Precision Bombing in the Post-Vietnam Era

After the war in Vietnam ended, Air Force planners refocused their efforts on preparing for direct conflict with the Soviet Union. In the subsequent decades the USAF shared, with its sister services, what one author has recently termed the "renaissance of American military power."[60] In the Air Force, this meant, principally, an increased emphasis on enhanced navigation techniques, and technologies producing further breakthroughs in precision guidance, low observables, infrared and radar sensors, and space-based systems. This process of reinvigoration began in the last years of Jimmy Carter's presidency, and continued with increased emphasis during the two-term presidency of Ronald Reagan.

Both the lessons of Vietnam and a sizeable expansion of the Soviet air forces between 1967 and 1977 prompted the USAF to emphasize not only the acquisition of a new generation of aircraft with advanced electronics, but also a program of robust and realistic combat training. Although the Boeing E-3 Airborne Warning and Control System (AWACS) met early resistance as an unpopular program, it quickly proved its worth as a long-range radar station and airborne command post. The AWACS aircraft, equipped with state-of-the-art communications, radar, and navigation systems, could be linked—during operations—to ships, ground stations, and other aircraft, allowing for the constant flow of vital information that had the potential to vastly improve target acquisition and accuracy.[61]

Although it had earlier roots in the British Mosquito bomber and the German "schnorkel" submarine, "stealth" technology (or "low observables") was spurred by the lessons of the 1973 Arab-Israeli War, which had demonstrated the increasing vulnerability of aircraft to radar-guided and heat-seeking missiles. While it does not make aircraft invisible, it makes them, as Richard Davis has pointed out, "difficult to detect and virtually impossible to track and engage."[62] The Stealth F-117 "Nighthawk" and B-2 "Spirit" aircraft are capable of flying undetected into enemy airspace and dropping bombs with the accuracy available from modern-day, laser-guided techniques.[63]

Even if the electro-optically guided bombs and laser-guided bombs of the late Vietnam period had begun a revolution in precision technology, the USAF adjusted to it only gradually. While it did upgrade its precision-guided munitions technology through the PAVEWAY series, it did not equip the majority of its newly acquired combat aircraft with that capability. In the mid-1990s, the USAF's entire precision-guided munition-capable fleet comprised only 125–135 fighter bombers.[64] Doctrine and operational guidance had not adjusted fully to the implications of all the newly available tools for air warfare.

In 1990 and 1991, the Persian Gulf crisis and subsequent war put American air power in the spotlight, testing its development in technology, target acquisition, command and control (including joint command operations), and logistics. The Gulf War also saw interesting modifications in Air Force bomber doctrine, influenced in part by Colonel John Warden's "Concentric Rings" targeting theory, which placed leadership in the center ring, followed by enemy industry and production, communications and transportation, the enemy population, and finally—on the outer ring— the enemy's armed forces. If Warden's views did not wholly upset past practice (the Air Force spent no small amount of time attacking enemy ground forces, bridges, and railways in 1991, for instance), they did at least challenge old orthodoxy on some levels.[65]

There were two distinct campaigns waged against Iraq by the U.S. military and its allies, under the aegis of the United Nations. In the Kuwaiti Theater of Operations (KTO), the Allies suppressed Iraqi air defenses, prepared the way for a Coalition ground attack, and supported ground forces with airlift and tactical firepower. In the strategic bombardment campaign against Iraq, the Coalition struck at a range of targets designed to erode the Iraqi will and capacity to fight, and to incapacitate the regime of Saddam Hussein. General H. Norman Schwarzkopf, head of Coalition Forces in the Persian Gulf War, would use the strategic air offensive as the first phase of an integrated campaign to liberate Kuwait. If the KTO absorbed the majority of the effort, the strategic campaign nonetheless received extensive attention from both planners and historians.[66]

At the beginning of the war in January 1991, the USAF had stationed one-quarter of its combat aircraft in Saudi Arabia, including 90 percent of its precision bomb-droppers. The U.S. Navy and Marine Corps, as well as Arab states and NATO partners, also contributed aircraft to the campaign.[67] Both new and old aircraft made important contributions to the air war over Iraq. B-52 bombers, constituting only 3 percent of total Coalition aircraft, delivered 30 percent of the total tonnage of air munitions. F-117 aircraft, carrying only two laser-guided bombs, achieved remarkable feats of accuracy. As Richard Davis has pointed out, they could "in some situations achieve the same degree of target destruction that in World War II had required 108 B-17s dropping 648 bombs."[68]

Coalition aircraft engaged in overwhelming assaults designed to destroy the Iraqi capacity to wage war, and will to wage war. Iraqi air defenses, communications and leadership targets, the electrical grid, oil refineries (and storage and distribution sites), and war-supporting infrastructure all came under simultaneous attack. Nuclear and chemical facilities also were hit by air, along with Iraqi airfields, bridges, and Scud missile launching sites. The Coalition dedicated 29 percent of its overall air campaign to the suppression or destruction of the Iraqi air defense system, the Iraqi air force, and the Iraqi navy.[69] Indeed, the air superiority and air supremacy that the Coalition quickly gained over the battle space made it difficult for the Iraqi ground forces to maneuver at will.

Attacks on communications and leadership targets consumed about 8 percent of total USAF effort in the Gulf War. There is no consensus on the real impact of these strikes: critics claim their impact was minimal, while Air Force advocates assert that they had an important role in degrading enemy effectiveness. What can be said, certainly, is that while air strikes did not paralyze enemy communications, Iraqi leaders nonetheless were forced to rely, often, on less efficient, less secure means of communication—particularly for troops in the field.[70]

In the space of days, Coalition air forces were able to shut down the southern and central Iraqi power grids, and to collapse the Iraqi oil-refining capacity. In part because of precision capacity, these targets could be degraded without massive damage; indeed, Iraq recovered much of its electrical generating power by mid-1992, and was exporting finished petroleum products by the end of that year. Precision and careful planning also helped to minimize Iraqi civilian casualties. The precision and target-finding capability available to the USAF in the early 1990s did not allow, however, for the rapid or complete elimination of the threat posed by Scud missiles; indeed, the mobile Scuds were largely immune to air attack at the time of the Gulf War. But the "Scud hunt" did help to suppress the Iraqi ability to launch the missiles—and it persuaded the Israelis to hold off on their own strikes against Iraq.[71]

Precision attacks on bridges complicated the ability of the Iraqis to resupply their troops and to move freely around the battle space. The latter contributed to traffic jams that made the Iraqis vulnerable to air action. Heavy attacks on ground forces undermined the will of Iraqi soldiers, convincing large numbers of them to desert rather than subject themselves to what seemed like a hopeless battle against a vastly superior foe. Much of the Republican Guard remained intact, however, despite a concerted Coalition air effort to destroy or degrade those units.[72]

The Gulf War air campaign has often been judged against the standard set by its original planners: to force the Iraqis out of Kuwait without a ground campaign, and possibly topple the Saddam Hussein regime in Baghdad. These goals, which were never accepted by the senior military leaders responsible for the campaign and thus were modified at the outset, set an artificially high standard against which to judge the success of the campaign. If its achievements were more modest, they nonetheless contributed in significant ways to the speed and success of the liberation of Kuwait.

The success of precision systems in the Gulf War and the constant march of advanced technology both contributed to the ongoing priority offered to precision in the 1990s. Ongoing conflict in the Balkans, fueled by the policies and actions of Serbian leader Slobodan Milosevic, drew the United States into the region repeatedly. In the autumn of 1994, NATO air forces struck, in a limited way, against Serbian forces fighting to recover Yugoslavia's breakaway region of Bosnia-Herzogovina. While an uneasy peace was brokered, NATO laid out plans for another air campaign that ultimately commenced in the late summer of 1995. Operation Deliberate Force stunned the Serbs, and probably contributed (along with victories won by the Croation-Muslim army in Western Bosnia) to a set of peace accords brokered in Dayton, Ohio. In this air campaign, the USAF relied heavily on precision-guided munitions. Over 200 aircraft flew more than 3,500 sorties, including 750 strike missions against Serb targets.[73]

Violence flared in the Balkans again in the spring of 1998 as Milosevic's repressive policies in Kosovo turned increasingly deadly. By the end of the year, it was clear that he planned a program of ethnic cleansing to establish complete Serb dominance in Kosovo. As last-ditch diplomatic efforts failed to sway the Serb leader from his course, NATO prepared to wage yet another air campaign in the region. Commencing in late March 1999, the campaign sought to use air power to coerce Milosevic into halting his trail of brutality and butchery. So optimistic were Western politicians about use of air power that they expected Milosevic to cave in to NATO demands within the space of a few days. This did not happen, and the Serb leader only intensified his program of ethnic cleansing. The NATO campaign, complicated by the differing aims and objectives of the many states involved, saw targeting restrictions lifted as Milosevic, in defiance of Western expectations, rode out air strikes and sought to outlast NATO's shaky resolve. Over time, the intensity of the air campaign increased and the target categories broadened. Both the Serb military, and the Serb civilian population came under intensified pressure, and in early June, in response to a deteriorating situation and, in particular, pressure from the Russians (who feared that a NATO ground offensive was increasingly likely), Milosevic announced that he would accept peace negotiations.

Precision munitions played a central role in the USAF campaign, which formed the bulk of the NATO campaign. The seemingly sudden capitulation of Milosevic prompted many air power supporters to argue that the campaign represented the first time that an air force, alone, had successfully defeated a ground force. But the precise reasons for Milosevic's capitulation have remained unclear. While the air campaign was surely a contributing factor, evidence indicates that other events—especially the threat of a NATO ground war, and the pressure placed on the Serbs by the Russians—were salient and central to the ultimate outcome. As Robert Pape has written, "As for Kosovo, we don't know whether Mr. Milosevic would have withdrawn his troops without the bombing of Serbia. However, while the bombs fell, American engineers were widening the roads in Albania, the British were calling up 30,000 reserves, and NATO countries had deployed 37,000 heavy ground forces to Kosovo's borders. This posed a credible ground threat that the Serbian leader could not ignore."[74]

The Global War on Terror

The September 2001 airliner attacks on the World Trade Center towers and the Pentagon (along with the failed attack that crashed in Pennsylvania), transformed the security environment in the United States, and opened the early stages of what the second Bush administration would term the "global war on terrorism." The first phase of that war commenced with attacks on members of the Afghan Taliban who had sheltered the al-Qaeda terrorists. Partnering with "Northern Alliance" opposition to the Taliban government, U.S. ground and air forces deployed an unlikely mix of ancient and modern tools against the Taliban—everything from mountain ponies to turbojet-powered "Global Hawk" drones. Moving quickly into Afghanistan after 9/11, the United States sought to rely principally on pairing its own precision-strike capabilities with the indigenous ground forces provided by the Northern Alliance. The Americans captured data on Taliban positions through satellite imagery, U-2 overflights, and drones. U.S. commandos carrying Global Positioning System receivers and laser range finders could fix the digital coordinates of enemy positions and call in air strikes. Approximately 60 percent of the bombs dropped in the campaign were laser-guided precision munitions. Air power was particularly effective against Taliban troops, especially when powerful Joint Direct Attack Munitions (JDAM) bombs were employed.

The role and effectiveness of air power in the Afghanistan campaign is still being debated by defense analysts, but several conclusions seem clear at this point. First, precision munitions were capable of doing swift and sometimes catastrophic damage to enemy positions. Second, new technologies like JDAMs added, in important ways, to the ability of the Americans to deliver powerful blows that undermined the enemy's ability to fight and will to fight. And third, precision-strike capabilities did not make other forms of military force unnecessary or obsolete in Afghanistan. Determined al-Qaeda fighters learned from early errors and found ways to elude detection or destruction by American air attack. Even the most sophisticated technologies cannot locate the enemy in all terrain and geographic conditions: al-Qaeda fighters were able to successfully utilize cover and concealment to evade the effects of American air power. They had to be met and overrun by traditional ground forces in battles at Bai Beche, Highway 4, and Operation Anaconda. And they were able to close, unseen, with friendly forces at Highway 4 and Sayed Slim Kalay.[75]

In early 2003, the USAF had another chance to demonstrate its precision capability when the United States attacked Iraq in order to topple the regime of Saddam Hussein. Unlike Operation Desert Storm of twelve years earlier, Operation Iraqi Freedom was designed principally as a ground campaign to overthrow the enemy regime quickly and decisively. Opening air strikes were carried out to prepare the way for the American advance and to impress upon the Iraqi leadership the overwhelming might of the American arsenal. These strikes, notable for their precision and pyrotechnics, neither eliminated Saddam Hussein (as many hoped they might) nor persuaded him to capitulate without any resistance. More notable and effective were the air strikes flown against Iraqi military and communications targets. Some of the most potent were flown early in the campaign while U.S. ground troops were halted in their tracks by blinding sandstorms. Far from being grounded due to bad weather, the USAF was now in a position to strike hard—and accurately—at the enemy under virtually all weather conditions. While the precise results of these strikes are still being assessed, there is no doubt that they severely degraded the Iraqi fighting capacity, such as it was, and advanced the speed with which the ground campaign could proceed once the storms subsided. The potent synergy of air power and ground power—utilized in mutual support and with a single aim—was proven in a decisive fashion.

Conclusion

The ability of the United States Air Force to hit any target it wants—with reliability and accuracy—has increased dramatically since the first half of the twentieth century. New technologies, undreamed of at the dawn of flight, have allowed bomber aircraft to achieve remarkable precision. The days of World War II–style area bombing are long past, never to return. As we move into the future, little seems beyond reach technologically: precision capabilities will only improve, and bombers will approach perfect accuracy rates. The ability to attack an enemy's assets and vulnerabilities by air—with precision and reliability—offers flexibility and dominance. It is a powerful and potent piece of the military toolkit that can be wielded by American policymakers.

Still, the triumph of precision bombing—achieved over long years of devoted effort—does not mean that wars involving U.S. forces will become simple and wholly predictable events. Even the most sophisticated sensing and intelligence-gathering devices cannot yet "see" in all terrain and under all circumstances: the surface of the earth is not yet transparent. This fact in itself will ensure that sophisticated enemies will continue to use cover and concealment with some degree of success, and will sometimes require ground forces to close in with and destroy them.

Moreover, finding and hitting targets—however accurately and reliably—does not itself equate to military victory. We still need to understand, as fully as possible, the relationship between dropping bombs and political outcomes. Intelligence—reliable, high-quality intelligence—is as important as it ever was: we need to understand the enemy we fight. The targets we judge to be central to the enemy war effort—or to the political coherence of the enemy regime—may not prove to be so important as we think. Just because we have new tools does not mean that the enemy cannot still confound us. Sensing technologies and satellite photos can only tell us so much; they do not offer the final word on the aims or intentions (or hidden capabilities) of our enemies. And room for resistance is often more available to an enemy than we expect. A regime that is not being forced to burn resources in a ground war will not be sensitive to the absence of those resources, and may be able to absorb tremendous amounts of punishment by transferring hardships largely to the civilian population. A dictator who is motivated to do so can attempt to use an entire nation as a human shield. This would raise ethical issues that might threaten the continued use of air power alone—however precise and discriminate it may have been.

Insurgency movements and civil wars still do not make themselves particularly amenable to precision air power. Often there are no targets so central to the war effort that their destruction would change the course of events. In addition, insurgents can control the timing of a war, waiting out a better-armed foe and resuming hostilities at the time—and under the circumstances—of their own choosing. And terrorists who prefer to use highly asymmetric methods of warfighting—often in conjunction with suicide delivery methods—will not themselves be deterred by the prospect of American air power, even if the regimes that harbor may well be.

Tools alone do not make strategy. A good strategy must be the foundation for every use of military force. But good tools offer immense advantages—and few military tools are so potent and so potentially useful as precision bombardment.

Notes

1. At this point, Tiverton was still serving with the naval section of the British aviation mission in Paris; see Tami Davis Biddle, *Rhetoric and Reality in Air Warfare: The Evolution of British and American Ideas about Strategic Bombing, 1914–1945* (Princeton, N.J.: Princeton University Press, 2002), 38–39.

2. For an excellent and comprehensive examination of Tiverton's work and his contribution to American thinking, see George K. Williams, "'The Shank of the Drill': Americans and Strategical Aviation in the Great War," *Journal of Strategic Studies* 19 (September 1996).

3. Biddle, 35–37.

4. Trenchard emphasized the "moral" effect of bombing particularly in his final war despatch. Anxious to claim that the resources devoted to bombing had been worth the effort, he argued that the "moral" effect was twenty times the physical effect of bombing. The complete text of the despatch can be found in AIR 6/19, Public Record Office, London. Parts of it were reprinted in H. A. Jones, *The War in the Air,* vol. 6 (Oxford: Oxford University Press, 1937), 136.

5. Williams, 396–99.

6. Biddle, 66–67.

7. For a detailed description of how the ideas evolved, see ibid., chap. 3.

8. Peter Faber, "Interwar U.S. Army Aviation and the Air Corps Tactical School," in Phillip S. Meilinger, ed., *The Paths of Heaven: The Evolution of Air Power Theory* (Maxwell AFB, Ala.: Air University Press, 1997).

9. Newton Baker, "Report of the Secretary of War," in *War Department Annual Reports, 1919,* vol. I (Washington, D.C.: Government Printing Office, 1920), esp. 68–75.

10. The notion of the Air Corps as an insurgency movement in the ranks of the Army comes from David E. Johnson, *Fast Tanks and Heavy Bombers: Innovation in the U.S. Army, 1917–1945* (Ithaca, N.Y.: Cornell University Press, 1998), esp. 81–90.

11. William C. Sherman, *Air Warfare* (New York: Ronald, 1926), 218.

12. Major Harold Lee George, among others at ACTS, spoke frequently of the vulnerabilities and interdependencies "which our present civilization has created." See "An Inquiry into the Subject 'War,'" lecture delivered at the Air Corps Tactical School [1934] in Air Force Historical Research Center (Maxwell AFB), decimal file no. 248.11-9.

13. Bruce Barton, "After Roosevelt—What?," *Colliers,* 21 Jan. 1939, 13, 35–36.

14. Steven L. McFarland, *America's Pursuit of Precision Bombing, 1910–1945* (Washington, D.C.: Smithsonian Institution Press, 1995), 94–98, quoted material on 95.

15. On tensions between the British and the Americans, see Biddle, 212–13.

16. See Churchill's 8 June 1941 comment on Portal's 3 June 1941 letter to the prime minister in Folder 2 (Prime Minister's Minutes, April–June 1941), Papers of Lord Portal of Hungerford, Christ Church, Oxford.

17. AIR 8/711, Public Record Office, London; Sir Charles Webster and Noble Frankland, *The Strategic Air Offensive Against Germany, 1939–1945,* 4 vols. (London: HMSO, 1961), 1:360–63.

18. "Target Priorities of the Eighth Air Force," Headquarters Eighth Air Force, 15 May 1945, Office of Air Force History, Bolling AFB, Washington, D.C., 520.317A; Stephen McFarland and Wesley Phillips Newton, *To Command the Sky* (Washington, D.C.: Smithsonian Institution Press, 1991), 96; Webster and Frankland, 2:27.

19. Webster and Frankland, 2:33, 36–37.

20. W. Hays Parks, "Precision and Area Bombing: Who did Which, and When?," *Journal of Strategic Studies* 18 (March 1995):148.

21. Richard G. Davis, *Carl A. Spaatz and the Air War in Europe* (Washington, D.C.: Center for Air Force History, 1993), 296–98; Richard G. Davis, "German Railyards and Cities: US Bombing Policy, 1944–1945," *Air Power History* 42 (Summer 1995):48–49.

22. Webster and Frankland, "Annexes and Appendices," 4:4.

23. For an informed and fascinating overview of navigational and bombing aids, see ibid., 4–15.

24. Biddle, 224.

25. Ibid., 228–29.

26. For a description of the methods and techniques used by the British and the Americans, see Webster and Frankland, (Appendices), Annex 1, "The Principal Radar Aids for Navigation and Bomb Aiming," 4:3–17, and Annex IV, "Bombs and Bombsights", 4:31–39.

27. Biddle, 229, 243–44; Parks, 147–58. For the text of the Butt Report, see Webster and Frankland, (Appendices), 4:205–13.

28. On marshaling yards raids, see Davis, "German Railyards and Cities"; on the Far Eastern air campaign, see Michael Sherry, *The Rise of American Air Power* (New Haven, Conn.: Yale University Press, 1987); Kenneth Werrell, *Blankets of Fire* (Washington, D.C.: Smithsonian Institution Press, 1996); and Conrad C. Crane, *Bombs, Cities and Civilians* (Lawrence: University of Kansas, 1993).

29. On intelligence and bombing, see F. H. Hinsley et al., *British Intelligence in the Second*

World War (London: HMSO, 1979), 1:3–43; Ralph Bennett, "Ten-Tenths Cloud Cover: Intelligence and Bomber Command," in *Behind the Battle: Intelligence in the War with Germany, 1939–1945* (London: Sinclair Stevenson, 1994); and Biddle, 276.

30. Biddle, 243.

31. Wesley Frank Craven and James Lea Cate, *The Army Air Forces in World War II* (Chicago: University of Chicago Press, 1951), 3:233–34.

32. On the air war in the Far East, see Michael Sherry, *The Rise of American Air Power* (New Haven, Conn.: Yale University Press, 1987); Crane, *Bombs, Cities and Civilians;* and Kenneth C. Werrell, *Blankets of Fire* (Washington, D.C.: Smithsonian Institution Press, 1996).

33. The USSBS was indeed so dismissive of British efforts that in 1951 the official historians of the American strategic bombing campaign felt compelled to apologize for its tone and conclusions. See Craven and Cate, 3:791. See also Parks, 145–47.

34. For a comprehensive history of these decisions, see the first volume of the official history of the Office of the Secretary of Defense: Steven Rearden, *The Formative Years, 1947–1950* (Washington, D.C.: Historical Office, Office of the Secretary of Defense, 1984).

35. David Rosenberg, "The Origins of Overkill: Nuclear Weapons and American Strategy, 1945–1960," in *Strategy and Nuclear Deterrence,* ed. Steven E. Miller (Princeton, N.J.: Princeton University Press, 1984), 122–23.

36. Ibid., 125

37. Ibid., 130–31.

38. The most authoritative book on the subject is Conrad C. Crane, *American Airpower Strategy in Korea, 1950–1953* (Lawrence: University Press of Kansas, 2000). A reliable summary of air power in the war can be found in Charles Gross, *American Military Aviation: The Indispensable Arm* (College Station: Texas A&M University Press, 2002), 150–72.

39. LeMay quoted in Thomas Hone, "Strategic Bombardment Constrained: Korea and Vietnam," in R. Cargill Hall, ed., *Case Studies in Strategic Bombardment* (Washington, D.C.: Air Force History and Museums Program, 1998), 517.

40. Crane, *American Airpower Strategy,* 90.

41. For an explanation of the Air Force philosophy, see Gross, 160–61.

42. Crane, *American Airpower Strategy,* 132–33.

43. Ibid., 90–91, 138–40.

44. Ibid., 138.

45. Ibid., 168–69.

46. Ibid., 159–63. See also idem, "Raiding the Beggar's Pantry: The Search for Airpower Strategy in the Korean War," *Journal of Military History* 63 (October 1999); and Biddle, 294–96.

47. Biddle, 296.

48. Between 1954 and 1957 the Air Force received an average share of 47 percent of total defense appropriations, compared to 29 percent for the Navy and 22 percent for the Army. See Rosenberg, 139.

49. On this period, see John T. Greenwood, "The Emergence of the Postwar Strategic Air Force, 1943–1953," in A. Hurley and R. Ehrhart, eds., *Air Power and Warfare* (Washington, D.C.: U.S. Air Force, 1979); Mark Clodfelter, *The Limits of Air Power* (New York: Free Press, 1989), 27; and Dennis M. Drew, "Air Theory, Air Force, and Low Intensity Conflict: A Short Journey to Confusion," in Phillip Meilinger, ed., *The Paths of Heaven: The Evolution of Air Power Theory* (Maxwell AFB, Ala.: Air University Press, 1997), 321–55.

50. On the free rein given to LeMay, see Rosenberg, 147

51. Quoted in ibid., 145.

52. See Steven Reardon, "U.S. Strategic Bombardment Doctrine Since 1945," in Hall, ed., *Case Studies,* 422.

53. Ibid., 423–32, quoted material on 424.

54. Drew, 328.

55. On the mismatch between Air Force doctrinal thinking and the reality of the Vietnam War, see Drew, "Air Theory," generally.

56. Hone, "Strategic Bombardment Constrained."

57. Earl Tilford, "Setup: Why and How the U.S. Air Force Lost in Vietnam," *Armed Forces and Society* 17 (1991):327. See also Robert Pape, "Coercive Air Power in the Vietnam War," *International Security* 15 (Fall 1990), and idem, *Bombing to Win: Air Power and Coercion in War* (Ithaca, N.Y.: Cornell University Press, 1996), 174–210.

58. Hone, 509–10; Richard Davis, "Strategic Bombing in the Gulf War," in Hall, ed., *Case Studies,* 529.

59. Hone, 498, 510, 514–16. See also, generally, William Momyer, *Air Power in Three Wars* (Washington, D.C.: Government Printing Office, 1978). On B-52s in Vietnam, see Thomas Keaney, *Strategic Bombers and Conventional Weapons: Airpower Options* (Washington, D.C.: National Defense University, 1984), esp. 20–23; and Davis, 529–36, esp. 531.

60. Gross, 218.

61. Ibid., 223, 226–27.

62. Davis, 533.

63. Gross, 232–33.

64. Davis, 529–30.

65. On Warden, see John Andreas Olson, "Colonel John A. Warden III: Smasher of Paradigms?," in Peter W. Gray and Sebastian Cox, eds., *Air Power Leadership: Theory and Practice* (London: Stationery Office, 2002); Lt. Col. David S. Fadok, "John Boyd and John Warden: Airpower's Quest for Strategic Paralysis," in Phillip Meilinger, ed., *The Paths of Heaven: The Evolution of Airpower Theory* (Maxwell AFB, Ala.: Air University Press, 1997); and Davis, 534, 540–57.

66. On air war in the Gulf War, see Richard Davis, "Strategic Bombardment in the Gulf War, in Hall, ed., *Case Studies;* Richard Davis, *Decisive Force: Strategic Bombing in the Gulf War* (Washington, D.C.: Air Force History and Museums Program, 1996); and Gross, 256–73.

67. Davis, 557–58.

68. Ibid., 573, quotation on 575.

69. Davis, 600.

70. Ibid., 590–93.

71. Ibid., 594–95, 598–99, 611. See also Gross, 266–67.

72. For a detailed explanation of the survival or Republican Guard units, see Davis, 605–11.

73. Gross, 280–81.

74. Robert Pape, "Wars Can't Be Won Only From Above," *New York Times,* 21 Mar. 2003, A23.

75. For the most detailed articulation of this argument, see Stephen Biddle, "Afghanistan and the Future of Warfare: Implications for Army and Defense Policy," Strategic Studies Institute, U.S. Army War College, November 2002.

Luftwaffe Intelligence: How It Viewed the United States Army Air Forces

James S. Corum

The *Luftwaffe* was the most formidable enemy the U.S. Air Force ever had to face in combat. Although greatly outnumbered in the air, the *Luftwaffe* was able to win some significant tactical victories in the Mediterranean theater. Over northern Europe, the U.S. bomber offensive of 1942–45 was bitterly opposed by the *Luftwaffe*. The *Luftwaffe* inflicted heavy losses on the Army Air Forces (AAF) throughout the air campaign and in late 1943 the German defense was so effective that the U.S. air commanders halted the raids deep into Germany until reinforcements and escort fighters could arrive.

Intelligence plays a key role in aerial warfare. What the Germans knew about the U.S. Army Air Forces (USAAF) helped determine the strategy and tactics used by the *Luftwaffe*. The intent of this article is to examine the *Luftwaffe*'s intelligence in the context of the air war against the Americans and to examine its strengths and weaknesses. Hopefully, a study of the huge intelligence effort devoted to the USAAF might provide some general insights into the nature of intelligence and operations and how air forces can effectively use, or misuse, intelligence. A useful starting point is to look at the German intelligence organization in general and the *Luftwaffe*'s place in it. The primary questions are fairly direct: what did the *Luftwaffe* know and when did they know it? Finally, I hope to come to an assessment of the general effectiveness of the *Luftwaffe*'s intelligence system and how it affected the war against the Americans.

German Intelligence About America in the Interwar Period

The German military put considerable effort into studying the U.S. military air arms and the U.S. aviation industry long before the *Luftwaffe* formally existed. After World War I, the German Imperial Air Service (*Luftstreitkräfte*), which had fought so brilliantly against the Allies and had pioneered an impressive array of new technologies and tactics for aerial warfare, was forced to disband under the provisions of the Versailles Treaty. Germany was forbidden an air force and even the civilian aviation establishment was limited by international agreement. However, Colonel General Hans von Seeckt, the visionary commander of the German Army from 1920 to 1926, was a strong believer in the importance of air power and was not going to see Germany deprived of a major weapon. Within the Reichswehr (German Army), a secret air force was created, complete with prototype aircraft development, secret training courses in Russia, and with a small but highly capable air staff manned by experienced veterans of the *Luftstreitkräfte.* The secret *Luftwaffe* staff was camouflaged in sections within the army staff.

The Reichswehr's senior army air officer from 1919 to 1927 was major, later lieutenant colonel, Helmuth Wilberg, one of Germany's most experienced airmen. He had joined the General Staff before the war and had learned to fly in 1910. During the war, he had ably commanded a force of over 750 aircraft during the great Flanders campaign of 1917–18. In 1919–20, he led an effort that involved hundreds of airmen to learn lessons from the war and to develop a practical doctrine for aerial warfare.[1] Looking for the time when Germany would rearm and build a new air force, Seeckt and Wilberg understood that air war doctrine could not, for long, be based on the experience of the World War. Air war doctrine and organization are closely linked to technology—to a degree even more than in ground warfare. For the secret *Luftwaffe* to remain current in air war and technology, it could not rely on its small-scale and very secret program of testing and training in Russia. The shadow *Luftwaffe* would have to rely heavily upon the experience of other nations. Although the United States Army Air Service had shown up in combat only in the last few weeks of the war, and even then flying mostly French and British machines, America had ended the war with a large aircraft industry and military air services. It was also clear that America was in position to become a leading civil and military air power. Consequently, even before the Nazi regime came to power, the Reichswehr and its shadow *Luftwaffe,* focused considerable effort on following American aviation developments, both civilian and military.

The Reichswehr intelligence office operated under the cover name of "T-3 Statistical Section"—called such so as not to offend Allied sensibilities with the word "intelligence." At least one aviation officer, assisted by some civilians, was assigned to T-3 during the 1920s to collect and analyze information on foreign air forces.[2] Although America was not seen as an immediate or even future threat to Germany, in the manner of Poland and France, the shadow air staff followed American developments very closely largely because America was the great power that was most accessible for German intelligence collection in the interwar period.

America, much more than Britain or France, welcomed visits and contacts with German officers and airmen after World War I. America had been in the war only a short time and had taken relatively low losses. In American eyes, the German Imperial Army and *Luftstreitkräfte* had fought an honorable war and had proven themselves superb fighters. Indeed, the prevailing view in the professional U.S. officer corps was that there was a lot to learn from the battlefield performance of the Germans.[3] Thus, when German officers began to visit in the United States in the early 1920s, they were allowed to visit U.S. Army posts, airfields, schools, and every manner of aviation facility with few restrictions. American officers were eager to talk with their erstwhile enemies, have them lecture on their war experiences and get their comments on U.S. training and maneuvers. Since Germany was not officially allowed to have military attachés before the early 1930s, visits by German officers, sometimes lasting three months or longer, replaced intelligence collection by attachés.

In one of his many reforms in German officer training and education, von Seeckt required all the officers in the Reichswehr to qualify with basic competence in a foreign language. For officers selected for general staff training, one had to earn a translator certificate in English, French, Russian, Polish, Czech, or Italian.[4] This level of language instruction required considerable fluency, and English seems to have been the most popular foreign language in the Reichswehr officer corps, with French a close second. Armed with people who could speak foreign languages, von Seeckt provided generous grants for officers to travel overseas for several weeks up to several months. While overseas, the officer was expected to take the opportunity to visit foreign armed forces, to collect open-source documents such as manuals and, upon his return, to write a detailed report for the Intelligence staff analyzing foreign doctrine, training, and technology. Several dozen of these reports from the 1920s still exist, and the United States was the most popular country for German officers, especially airmen, to visit.[5]

The surviving reports provide a fairly comprehensive picture of the U.S. Army Air Corps in the 1920s and early 1930s. Major Wilberg, chief of the Reichswehr's air staff, spent several weeks visiting U.S. Army Air Service installations in 1925–26 along with Captain Bauemkler, one of his top technical experts.[6] Major von dem Hagen, the senior air officer on the Reichswehr Intelligence Staff, visited the United States in 1928 and wrote an extensive report on the Army Air Corps.[7] The visit of Captain Speidel of the Reichswehr air staff to the United States in 1929 illustrates the friendly reception given to the veteran German airmen a decade after the World War. He stayed in the United States for several months and spent two to six weeks at the Air Corps Tactical School, the Engineering School, the Technical School, and the Primary and Advanced Flying Schools. He visited observation, attack, bomber, and fighter squadrons and observed their operations. He sent several boxes of U.S. Air Corps manuals back to the Reichswehr Intelligence Staff. His generous Air Corps hosts even let him fly several of the latest army aircraft.[8] As with most of the German airmen who visited the United States, he was generally impressed with the organization and the equipment of the Army Air Corps. Although the Air Corps was fairly small, the Germans were impressed by its excellent training establishment and the competence and professionalism of its flying officers. Moreover, the Army's aircraft of the late 1920s and early 1930s were on par with the best military aircraft of the time, and the visits gave the Germans a good idea of the recent developments in aviation technology.

Indeed, American aviation writing and doctrine was carefully and critically analyzed throughout the interwar period, and this information was published and disseminated through the German military by the Reichswehr Air Staff. The Reichswehr air staff subscribed to the major foreign air journals, including the primary American journals. Starting in 1919, the Reichswehr air staff published a regular newsletter for officers called "Air News" (*Luftfahrtnachrichten*) which emphasized foreign technical developments.[9] The major journal of the German Army, the *Militär Wochenblatt,* covered foreign air power developments and thought in considerable detail. The

Germans commented with considerable and critical insight on General Billy Mitchell's writings and Royal Air Force (RAF) Air Marshal Hugh Trenchard's speeches, as well as the latest published doctrine of the British, American, and French air forces. In short, despite the small size of the shadow *Luftwaffe* (about 500 airmen in 1933 when the Nazis came to power) the German airmen developed a comprehensive picture of the U.S. Army Air Corps and of the state of American aviation technology. Indeed, by the early 1930s, it is likely that the Germans knew more about American aviation than they did about the British or French air forces that were considerably less welcoming to the German officer visits. The tone of the German reports and comments on the Americans was one of respect. While the American military aviation establishment was small, it was seen as on the cutting edge of aviation technology and theory. What the Weimar Germans saw in American aviation was a great and perhaps very dangerous potential, especially in aircraft production and engine technology.

The Nazis Assume Power—Expansion of Intelligence

When Adolf Hitler came to power in January 1933, he had at his disposal the small but capable intelligence sections of the army and navy and specialists in the Foreign Ministry experienced in collecting intelligence and employing ciphers. Hitler was a devoted consumer of intelligence, and, by the middle of the 1930s, a large intelligence bureaucracy had grown from small beginnings.

The story of German intelligence in World War II is a very complex one, but it must first be noted that the Nazi state was loaded with an enormous burden of bureaucracy—so much so that it became a matter of many large agencies collecting and analyzing intelligence data, sometimes working in parallel and sometimes at cross purposes. Without going into a detailed overview of the German intelligence organization, it should be noted that the Third Reich had at least seven major civilian intelligence agencies and the Wehrmacht several more. The Reichspost (Postal and Telephone Agency) had a section devoted to deciphering foreign diplomatic codes and message traffic, the Nazi party had a large department for monitoring and controlling the activities of German nationals abroad and for employing German citizens and Nazi sympathizers in espionage and subversion networks, the Economics Ministry had an office for collecting information about foreign economic information, Josef Goebbels's Propaganda Ministry had a department for monitoring foreign press and events, the SS—as might be expected—had several sections devoted to both internal security and collecting foreign intelligence. Within the armed forces, the Abwehr (Foreign Intelligence Service) under Admiral Wilhelm Canaris was set up in 1935 with the task of running espionage and counterespionage operations. It was loosely under the direction of the Oberkommando der Wehrmacht. The army, navy and *Luftwaffe* all had their own intelligence organizations and staffs. In addition, when the Reichs War Ministry set up a Military Economics Branch to plan and coordinate armaments production in 1934, it contained a research section to collect information of foreign armaments and production to provide strategic insight into the likely requirements for German strategic planning.[10] In addition, there were dozens of academic and industry research institutes that collected information of foreign industry and technology and carried out special studies for one of the military or civilian intelligence agencies.

Despite a vast apparatus for collecting intelligence, the Third Reich had no organization tasked with the responsibility to coordinate intelligence collection or to provide a central point for a comprehensive evaluation of the vast quantity of data that flowed in. The branches of the armed forces willingly shared information through liaison officers and normal staff contacts, but the cooperation was usually informal and often haphazard. There was no reliable system or method to sift through the data and provide an accurate picture of foreign capabilities and intentions to the top leadership of the Third Reich. Of course, if Hitler had had such an intelligence organization to provide strategic intelligence, it probably would not have lasted long. Hitler had little use for strategic analysis that contradicted his own genius and wanted all the major intelligence agencies to report directly to him. The Führer preferred to be his own intelligence analyst, free to latch on to whatever data that supported his strategy and disregard the rest.

Prior to the outset of the war, Hitler devoted little attention to America and to the American military. Hitler's focus was on Europe, and the small armed forces of an isolationist America posed little threat to his immediate political goals. Even when the war broke out in Europe, Hitler and his inner circle were interested only in the ability of the Americans and American industrial production to influence European events in the short term. In the early stages of the war, America simply did not figure into the Führer's grand strategy. However, his lack of interest was not shared by the German armed forces, which continued to collect a large quantity of intelligence on the United States, its armed forces, and especially its military aviation.

German Strategic Intelligence on American Aviation

From 1933 to 1941, Germany's primary intelligence source on the U.S. Army Air Corps was its military attaché in America, Major General Friedrich von Boetticher. For fourteen years after the World War, the Allied powers had denied Germany the right to send accredited military attachés abroad, but by the early 1930s this policy was seen as needlessly humiliating for the Germans and a barrier to better diplomatic and economic relations. In early 1933, the attaché system was reintroduced, and the major powers announced that they would be pleased to receive and accredit German military attachés. The officer selected to represent the German military in America was exceptionally qualified for the post. Von Boetticher had joined the general staff before World War I and during the war compiled a solid record as a staff officer. Well educated and well read, von Boetticher spoke perfect English. From 1920 to 1924, von Boetticher served as chief of intelligence in the Truppenamt (camouflaged general staff) and showed a knack for intelligence collection and analysis. He moved on to eventually command the army artillery school, a very prestigious post, and was promoted to major general in 1930. He also had the advantage of knowing America well. He spent five months in the United States in 1922 visiting U.S. Army posts, made some good contacts in the U.S. Army, and gained the confidence of American officers by being willing to share information on the German Army. On his return to Germany, he sent the Army War College some detailed reports on how the German Army had managed railroad logistics during the war. He was a perceptive and personable man with genuine respect for Americans and a very strong interest in American military history. In short, he was almost ideally suited to attaché work.[11]

Von Boetticher sent his reports on the Americans to the War Ministry where they were circulated among the senior officers and staffs. He had an abiding interest in military aviation and spent a great part of his time observing and reporting on the state of the Army Air Corps and the U.S. aviation industry. For example, in 1933, he visited the Air Corps's testing and research center at Wright Field and also tactical units at Langley Field. In addition, he toured the Pratt and Whitney engine factory.[12] One of the first American officers he got to know well was General Benny Foulois, chief of the Army Air Corps. In late 1933, von Boetticher reported to Berlin Foulois' comments that the long-range bombers would soon outperform fighters in both range and speed—a vision that became reality when the Army Air Corps acquired the B-17 bomber in 1935.[13] Von Boetticher visited army airfields, observed maneuvers and even managed to get rides in U.S. Army aircraft. His reports consistently described American aviation equipment as very good, and he sent large amounts of open-source information on the U.S. aircraft industry to Germany. In his report on the U.S. Army in 1937, he commented favorably on Air Corps equipment, tactics, and training.[14]

One of von Boetticher's routine jobs as attaché was to clear visits of German scientists and engineers to U.S. factories and installations with the War Department. Between 1933 and 1938, over 400 German aeronautical engineers and designers visited U.S. aircraft manufacturing and test facilities. These included senior aviation experts from the German Aviation Testing Center.[15] Until relations with the Germans became strained in 1938, German aviation experts had fairly open access to American aviation technology and were glad to use open-source American information in furthering the rise of the German aircraft and engine industries. In interrogations conducted in June 1945, German aircraft designers told AAF analysts how they had used the excellent and openly published National Advisory Committee for Aeronautics (NACA) wind-tunnel data to design propellers for German warplanes. The NACA wind tunnels were described by knowledgeable Germans as better than their own.[16] If Hitler and Hermann Goering were slow to appreciate the capabilities of American technology and production, there were hundreds of German aviation specialists who knew a great deal about American aviation. However, as capable an observer as he was, von Boetticher was not an aviator. So, in 1938, the *Luftwaffe* sent Peter Riedel to assist von Boetticher in collecting and evaluating information on American aviation. Riedel was a surprising pick as air attaché, as he had no military experience. He was, however, a highly accomplished aviator with several records to his name and experience as a test and airline pilot. Moreover, Riedel had a degree in aeronautical engineering and could give von Boetticher's reports on U.S. aviation (which were jointly written) some extra credibility.[17]

As Europe lurched toward war in 1939, Hitler had only one question about the Americans: how soon could they provide support to Britain and France in case of war? Riedel helped von Boetticher prepare estimates for Hitler of U.S. aircraft production and the strength of the Air Corps from 1939 to 1941. The estimates of American production and strength were fairly accurate. In early 1940, Von Boetticher and Riedel stated that the U.S. aviation industry was just beginning mobilization and could not provide much help to Europe that year—a maximum of 1,250 bombers and 900 fighters. In fact, this overestimated U.S. capability at the time.[18] In May 1940, commenting on President Franklin D. Roosevelt's call to boost U.S. aircraft production to 50,000 planes a year, the attaché argued that United States would be in no position to influence anything in Europe in the short term. He doubted that the 50,000-plane figure was realistic but repeated his position that the Americans had tremendous industrial potential.[19]

Through 1941, as he accurately reported America's unreadiness for war, von Boetticher hastened to warn that his assessment of the Americans applied only in the short term. In a cable to Berlin in July 1941, he noted that America was rapidly mobilizing its industry and military forces and that the U.S. Army had grown to over 1.4 million men, and that the regular army and the Air Corps would be fully equipped by the spring of 1942. With a fair idea of how the Nazi inner circle thought, he added the following warning about his own reports:

> In my reports I have regularly noted the development of American armaments and the armament industry, also their weaknesses. I urgently warn against overestimating the weaknesses and underestimating American efficiency and the American determination to perform. . . . As I have done for years, I repeat in particular my report that the American officer corps of the Army and Air Corps in general meets high requirements. . . . They are giving the greatest attention to the problems of modern warfare.[20]

As von Boetticher suspected, his generally sound assessments of American military and air power were generally ignored by Hitler and the Nazi inner circle as they developed their war strategy. Reichsmarschall Hermann Goering, commander in chief of the *Luftwaffe,* generally expressed contempt for American industry before 1943. In 1939, he remarked to the *Luftwaffe*'s intelligence chief, General Beppo Schmid: "The only thing they (the Americans) are good at is making automobiles—but not planes." Even in January 1942, he assured Benito Mussolini that "America is all talk and no action," but then added that "If the war lasts much longer we must assume that the Axis is going to feel something of the planes being produced by America."[21] Sometimes Hitler even recognized the American threat. In 1940, Hitler admitted that even if von Boetticher's reports of American military and air potential were accurate—it did not matter as he expected that by 1942 Britain would be knocked out of the war and Germany would dominate Europe.[22]

When von Boetticher was repatriated back to Germany with the other diplomats in early 1942, he was given a brusque reception by Hitler who thanked him, presented him with a medal for his service, promoted him to lieutenant general and immediately placed in on the "Leaders Reserve" to await the call for a new assignment—a call that never came. Von Boetticher had not made himself popular in Berlin with his pessimistic reports about American military and industrial potential. For the rest of the war, this highly experienced general staff officer in excellent health lived in Potsdam and occupied his time with writing intelligence analyses of American industry, doctrine, operations, and intentions. His intelligence analyses were, as usual, quite perceptive. His reports had a receptive audience in Oberkommando der Wehrmacht where they were read by General Wilhelm Keitel and his papers quietly circulated among senior military leaders.

In time, many of the senior officers came around to Von Boetticher's assessment of America's warmaking potential. But by then it was far too late to rectify the great strategic mistake made by Hitler early in the war—that the war would be short and Germany did not need to completely mobilize its war industries and training programs to contend with American industrial might. Although some of Germany's top aviation experts, such as Franz Siebel, implored the Reichs Air Ministry and Hitler in 1940 to take the American production potential seriously and begin a rapid buildup of German airplane production, Hitler ignored all such advice.[23] As late as 1943, Hitler would argue that the American airplane production figure of 50,000 planes in 1942 was grossly exaggerated. One advantage of having a large number of uncoordinated intelligence agencies is that a determined leader can shop among them to find intelligence data that fits his preconceived notions. This is precisely what Hitler did in 1943 when he quoted estimates by the Economics Ministry intelligence section to prove that America did not have the raw materials to build as many planes as von Boetticher and some military analysts argued. Only by ignoring the growing aerial threat to the Reich in 1943, which would cause a diversion of resources to the *Luftwaffe,* could Hitler continue with his plans for the great Russian summer campaign that would climax at Kursk.[24]

The refusal of Hitler and Goering to take the American threat as well as the Russian war potential seriously meant that not until 1943 could Field Marshal Erhard Milch, state secretary for aviation, convince the Führer and High Command to take measures to dramatically increase fighter production for homeland defense. The need to increase pilot and aircrew numbers was also ignored. Only in late 1942 did the *Luftwaffe* revise its fairly small pilot training program to produce the numbers of pilots that would be needed to defend the Reich from the Americans. By then, it was too late. The only way that the pilot numbers could be increased was to drastically reduce the amount of flight time for trainees. By 1943, the new fighter pilots assigned to Reich air defense had barely 130 hours of total flight time, as opposed to the American and British pilots with over 400 hours.[25] As the air war turned to attrition, the inability of German pilots to survive in combat against better-equipped and better-trained Americans became the crucial factor in losing control of Germany's airspace.

The *Luftwaffe* Intelligence Organization

When the *Luftwaffe* was officially created in March 1935, it took over the shadow *Luftwaffe* intelligence staff as well as a cadre of specialists in signals, reconnaissance, and aerial photography. Within the *Luftwaffe* general staff, the old Reichswehr branch of T-3 (Air Intelligence) was renamed 5th Branch (Foreign Air Forces). The General Staff Intelligence section was further divided into eight sections. Section 1 was the command section, Sections 2 and 3 dealt with the Western air forces, Section 4 dealt with the USSR and Balkans, Section 5 was photo interpretation, Section 6 was the press office, Section 7 managed the military attachés, and Section 8 troop welfare. In addition, there was a target office that prepared target folders for the air staff.[26] The *Luftwaffe* general staff also had two other agencies directly subordinate to it, a central photo interpretation command and a special squadron of long-range photoreconnaissance aircraft. The *Luftwaffe*'s special reconnaissance squadron was composed of civilian aircraft and airliners with aerial cameras built in and carefully hidden. Before the war, these aircraft flew over the breadth of Europe photographing potential war targets. Often the German airliners would "accidentally" stray off course to obtain photos of military installations which ended up in the target folders of the *Luftwaffe* Intelligence staff. Early in the war, the *Luftwaffe*'s 5th Branch formed a special technical intelligence center at the large German airbase and test center at Rechlin where captured enemy equipment was analyzed. When possible, damaged enemy aircraft were repaired and flown by *Luftwaffe* test pilots and even operational pilots to give *Luftwaffe* commanders a better "feel" for enemy equipment and capabilities.

Another organization directly under the *Luftwaffe* intelligence staff formed at the outbreak of the war was a special interrogation center where downed enemy airmen would be held for up to twenty-five days in solitary confinement, where a group of expert interrogators would try to pry information out of their captives. When the interrogators had exhausted the intelligence value of the downed airman, he was sent on to a regular *Luftwaffe* prisoner-of-war camp to sit out the rest of the war. The *Luftwaffe*'s interrogation center, called Dulag Luft (Dulag—short for *Durchgangslager* or "transit camp") and located in Oberursel near Frankfurt, was such a good source of intelligence that a few small farm buildings soon expanded into a large complex that processed thousands of Allied airmen. Dulag Luft also controlled two smaller interrogation centers, one in Verona for Allied airmen shot down in North Africa and Italy and one in Wetzlar for captured Russian airmen.

The *Luftwaffe* general staff was organized into branch inspectorates for bombers, fighters, aircraft armament, and so on. The branch inspectorates were responsible for the doctrine, training, and equipment development of their respective branches. The branch inspectorate for reconnaissance, which was responsible for the *Luftwaffe*'s long- and short-range reconnaissance squadrons, worked closely with the intelligence staff to meet the planned requirements of the operational commands and of the *Luftwaffe* staff. The reconnaissance units of the *Luftwaffe* were assigned to support the various air fleets and air corps of the *Luftwaffe* but the *Luftwaffe* general staff could direct the air fleets to use their reconnaissance assets to collect specific information. The *Luftstreitkräfte* of World War I had developed photoreconnaissance to a fine art, and this tradition was carried on into the new *Luftwaffe* of the 1930s. The Germans started the war with superior aerial cameras, and the air fleets and air corps had small photo interpretation detachments to provide high-quality aerial intelligence to the *Luftwaffe* and army commanders and to the *Luftwaffe* intelligence branch.

The largest and most important intelligence collecting organization of the *Luftwaffe* was the Signals Branch (*Luftnachrichtentruppen*). The Signals Branch was responsible for all *Luftwaffe* communications, radar units, air controllers, ground observers, and electronic warfare units. The Signals Branch came to employ tens of thousands of soldiers organized into signal regiments, battalions, and companies. Of especial importance for intelligence collecting were three signal regiments that specialized in intercepting and analyzing Allied radio traffic. These were Regiment 353 in Russia, Regiment 352 in the Mediterranean, and Regiment 351 that specialized in the Allied air forces in the West. Each Regiment contained highly specialized battalions, such as the 357th Signal Battalion of the 351st Signal Regiment, which monitored Allied heavy-bomber radio traffic.[27] The battalions employed well-trained linguists, many of them women, to monitor and translate the radio conversations of Allied ground and air units. The German radar and air controllers were, of course, the primary intelligence collectors of the *Luftwaffe*. The radar units reported the flight of Allied aircraft over Germany and German-held territory, and the air controllers would assess the routes and probable targets of the Allied planes. Flak, fighter units, and civil defense forces were given warning, and the regional or air fleet command centers would orchestrate the German defense to inflict maximum losses upon the enemy.

The intelligence branch also assigned liaison officers to the major headquarters of the army and navy staffs. The *Luftwaffe,* unlike the RAF and AAF, did not have intelligence officers at the group or wing level. *Luftwaffe* battle reports and reports on enemy equipment were handled by the group and wing operations staffs and passed on to higher headquarters. The *Luftwaffe* has been criticized by many air power historians for having a low regard for intelligence, based on evidence that few trained general staff officers or regular officers were ever assigned to intelligence duty.[28] Almost all of the intelligence officers assigned to the air fleets and air corps were reservists from varied backgrounds who were expected to learn "on the job." However, this criticism of the Germans falls flat when one notes that the British and Americans were not very different in their attitude toward filling intelligence slots. The most promising regular officers were placed in troop command or in operations—not in intelligence. General Omar Bradley remarked on the U.S. view of intelligence assignments from the old regular army perspective: "Misfits frequently found themselves assigned to intelligence duties. And in some stations G-2 became a dumping ground for officers ill-suited to command, I recall how scrupulously I avoided the branding that came with an intelligence assignment in my own career."[29] In fact, many of the American, British, and German reserve officers thrown into intelligence duties during the war proved to be quite good at the work. In looking at the performance of the Allied and Axis armies and air forces. it appears that some officers had an inborn knack for intelligence work—despite coming from the world of academia or business. Indeed, it would be difficult to make a case that intelligence collection or analysis suffered from a lack of regular officers in those duties. It can even be argued that soldiers from outside the traditional military career system might be better suited to intelligence, work that requires considerable creative and analytical skills, than someone trained in the more rigid and bureaucratic culture of the regular forces.

The *Luftwaffe* general staff's chief of intelligence from 1938 to 1942 was Colonel, later Major General, Josef "Beppo" Schmid. The hard-drinking Schmid, known for his reputation as a playboy, has usually been portrayed as an incompetent intelligence officer, way out of his depth. He won this reputation fairly for his incredibly inaccurate assessment of the RAF's Fighter Command strength and capabilities during the Battle of Britain.[30] Indeed, it was his worst hour. However, I suspect that Schmid may not have been as bad as usually portrayed. Under his direction, *Luftwaffe* intelligence performed creditably in the battles for Poland, Norway, and France.[31] It was in the nature of the Nazi High Command to ignore or reject pessimistic intelligence analysis and more than a few senior officers who persisted in bringing unhappy news to Goering or the Führer found their careers quickly cut short and exiled to minor posts. Schmid had to survive in the higher echelons of the staff where bad news had to be carefully tailored and minimized to be palatable. Schmid somewhat redeemed himself as an intelligence officer, and consequently made himself unpopular with Goering and Hitler, by accurately describing the renaissance of the Red Air Force and Soviet aircraft production through 1942 as well as warning of the potential American bomber threat. Schmid's situation reports, called *Lagenberichte* in German, were soon ridiculed and punned as *Lügenberichte*, or "Lies Reports," by Goering.[32] In October 1942, Schmid was ostensibly removed because a *Luftwaffe* lieutenant in the attaché section was uncovered as a Russian spy.[33] However, the point was also clearly reinforced that bad news was not welcomed at the top.

Schmid's replacement, Colonel Rudolf Wodarg, served as *Luftwaffe* intelligence chief from March 1943 to 1945. By all accounts, he was a very talented and capable intelligence chief. However, he also had to adopt the, by then, standard general staff practice of feeding unfavorable intelligence to Goering and Hitler in small doses. Even then, he faced the ire of Goering for reporting such facts as the increasing range of U.S. escort fighters. By 1944, and especially after the 20 July plot, intelligence implying Germany's increasingly hopeless situation in the air was liable to be labeled "defeatist," and the senior staff officers subject not only to relief but investigation. Given the poisonous atmosphere in the OKW (Oberkommando der Wehrmacht—Wehrmacht High Command) briefings in Berlin, in order to present a comprehensive analysis of the American air strategy and threat, Wodarg had to resort to forging documents, ostensibly from General Carl A. Spaatz, commander of the USAAF strate-

gic forces in Europe. In January 1945, Wodarg presented two mud-spattered documents, dated 5 and 24 January, in English under Spaatz's name and purporting to be
top-secret directives outlining American air strategy. As raw strategic intelligence,
Wodarg could simply vouch for their authenticity and pass them on to the High
Command. In reality, Wodarg and his deputy had written the directive.

The forged documents provide a good picture of *Luftwaffe* intelligence analysis
of USAAF strategy. The documents are decent forgeries as far as format is concerned. While the authors failed to get Spaatz's writing style or directive form quite
right, USAAF analysts after the war said the forgeries, especially the content, were
"pretty plausible." The directive of 5 January predicted that the AAF bombers would
focus their primary attention upon paralyzing the German rail and water communications system—anticipating "Operation Clarion" as well as a continued major campaign against bottleneck targets in the chemical and oil industry. The directives
called for a continued bombing effort of over 100,000 tons of bombs per month and
foresaw the attack on Dresden by calling for "attacks on the Eastern German communications system" to support the Soviet offensive.[34] Essentially, Colonel Wodarg's
analysis of the AAF strategy was on target.[35] It merely illustrates the extreme measures competent intelligence officers had to take to get sound information to the top
decision-makers.

Operational Intelligence—Combat with the AAF in the Mediterranean Campaign, 1942–1943

Although a few American bombing groups based in England carried out some
missions against targets in France in mid-1942, the first large-scale clash between the
Luftwaffe and American Army Air Forces came in North Africa in late 1942. The Germans beat the Americans in the race for Tunisia, and in late 1942 to early 1943 easily
held the Americans at bay. In the early weeks of the campaign, the Germans held air
superiority over the Tunisian Front and Goering seems to have gained a false impression of the American air arm. The American problem was not due to deficient leadership, training, equipment, or numbers. At the start of the campaign, the AAF had to
operate at the end of a very long supply line from small, unimproved forward airfields
that turned to mud in the winter rain. It took time to sort out the numerous logistics
problems and get adequate fuel and spare parts to the front. Engineers had to be
brought up to build adequate airfields. At first, the Americans simply could not get
enough fighters and light bombers into the air to provide air support for the ground
forces and to interdict the Germans. In the meantime, the *Luftwaffe* had the advantage
of flying from good all-weather airfields captured in Tunisia.

However, the German advantage was soon gone. By spring, a reorganized AAF,
with its airfield and supply problems sorted out, easily smashed the *Luftwaffe* in the
air while providing massive air support to the forward troops and interdicting a great
part of German air and sea transport. The Allied air superiority over Tunisia from
March to May 1943 played a decisive part in forcing the surrender of over 250,000

Axis troops. Long before the Axis collapse, the German and Italian air forces had been run out of the sky by vastly superior American and British air forces.

The German high command, stunned by the catastrophe in Tunisia, hastened to reinforce Italy with air and ground forces. The *Luftwaffe* also sent some of its best tactical commanders to meet the American and British air threat in the Mediterranean. In June 1943, Field Marshal Albert Kesselring, commander of German ground and air forces in the Mediterranean, relinquished command of the 2nd Air Fleet (while retaining the title of theater commander and command of German ground units) to Field Marshal Wolfram von Richthofen, who had just left command of the *Luftwaffe* forces in southern Russia. On his way from Russia to Italy, Von Richthofen called on Goering to discuss air strategy. Goering told von Richthofen that the *Luftwaffe*'s failure in North Africa had not been attributable to any Allied superiority but to cowardice on the part of the *Luftwaffe*'s pilots. Goering issued an order authorizing him to relieve any pilot or *Luftwaffe* officer, up to wing commanders, who showed a lack of aggressiveness before the Allied airmen. These "cowardly" officers would be sent to the Russian Front to fight as infantrymen.[36]

When von Richthofen arrived in Italy in June, he gleaned a very different picture of the American and British air forces. The 2nd Air Fleet staff calculated that the American and British air forces outnumbered them by a factor of 4–5 to one. Moreover, the 2nd Air Fleet also maintained that the Americans and British were also superior in the realm of aircraft technology. The American heavy bombers in North Africa had the range to attack any German airfield in Italy or southern France, while the *Luftwaffe*'s small force of Ju 88 bombers could reach only a fraction of the Allied air bases. The Allied medium bombers, such as the B-25s and B-26s, were considered to be equal or superior to the German bombers. The new Allied fighters, especially the latest Spitfire models (which equipped some of the U.S. fighter squadrons) and the P-38s and P-47s, were considered to be superior to the Me 109 in most respects. Because of Allied superiority, the old Ju 87 Stukas, the *Luftwaffe*'s best anti-ship weapon of the time, could not be easily employed. In addition to their superiority in numbers and equipment quality, the Allied pilots were rated as highly trained; so the *Luftwaffe* could consider itself open to Allied attack at all points and in bad weather. Aside from singling out the capabilities of the U.S. bombers, the *Luftwaffe* made little distinction in rating the effectiveness of the British and American fighter forces. Both were highly trained, well equipped and supported by an excellent logistics network. Given the estimate, which reflected a very sound general appreciation of the dilemma faced by German air power, Von Richthofen had to quickly develop tactics to deal with the Allied aerial superiority.

The Americans and British carried out heavy air attacks against the German and Italian air forces throughout June and early July 1943. There were also strong attacks against the rail centers throughout Italy and some long-range attacks against Italian war industries in the North. The problem facing the Germans in the Mediterranean was to try to predict what move the Allies would make next. Von Richthofen and the 2nd Air Fleet staff had few long-range intelligence assets capable of monitoring

Allied activity. One of the best German assets in the theater, the signals monitoring stations, were hampered by the fairly strict Allied radio silence directed in the few weeks before the Allied landings in Sicily. The *Luftwaffe,* which had not recovered from its losses in the Tunisian campaign, had only one long-range reconnaissance squadron (Ju 88s) for the entire Mediterranean theater. In the face of Allied air superiority and air defenses around the major ports and staging areas in North Africa, German aerial reconnaissance was spotty indeed. Only occasionally could the *Luftwaffe* slip a plane in past the watchful Allied defenses to photograph the ports, bases, and airfields. For covering the open ocean and Italian coastline, the Germans and Italians had a few seaplanes and other short-range reconnaissance aircraft to search out Allied convoys.

In 1942, the *Luftwaffe*'s aerial reconnaissance in the Mediterranean and over Britain had enjoyed a brief renaissance with the development of a few specially adapted Ju 86 bombers which could fly at very high altitudes (over 42,000 feet) and photograph Allied bases safely above the effective ceiling of Allied fighters and anti-aircraft.[37] However, the British soon countered this tactic with a unit of Spitfires that were also specially adapted to fly at high altitudes and intercept the German intruders. After a couple high-altitude reconnaissance planes were shot down, the *Luftwaffe* pulled the Ju 86s out of combat and the Germans were faced with the dilemma of using their Ju 88s on extremely risky missions.[38]

For the *Luftwaffe* in the Mediterranean, the strategic intelligence provided from the OKW's Abwehr and from Berlin was worse than useless. In the weeks before the invasion of Sicily, German agents in North Africa provided some appallingly inaccurate data to the Abwehr and senior staffs. German agents estimated that the British, Americans, and Free French had approximately fifty-five divisions in the theater, mostly in Tunisia and Algeria, and that half of them were fully trained and equipped for combat.[39] In fact, the Allies had less than half of this figure. Moreover, other bits of exaggeration slipped into the German estimate of the Allied Table of Organization and Equipment. For example, a German agent reported three U.S. paratroop divisions in the theater, all stationed near Oran.[40] In fact, the Americans had only one paratroop division (82nd Airborne) in the theater. Although British counterintelligence had managed to find and "turn" most of the German agents in Britain and use them to feed false information to the Germans, the bad intelligence provided by the German agents in the Mediterranean does not appear to be the product of a clever Allied deception campaign. German agents really were that bad. As the war progressed, the intelligence provided by Abwehr agents came to be rightly viewed by the German staffs as being generally unreliable.[41]

As happened on more than a few occasions during the war, the Wehrmacht and *Luftwaffe* staffs in Italy were largely on their own in developing an accurate intelligence picture for their theater. Despite a lack of signals information and aerial reconnaissance, Kesselring and von Richthofen used their experience and common sense to predict the Allied moves with considerable accuracy. In June, the increased tempo of Allied attacks on the German airfields in southern Italy and of vital rail lines in the peninsula clearly indicated that an attack was coming. The army and *Luftwaffe* intelligence staffs in Italy believed that the Allies had enough shipping to land twelve ground divisions in Sicily—a fairly accurate assessment. Even before Tunisia fell, Kesselring and the 2nd Air Fleet emplaced long-range Freya radars in Western Sicily and another in Sardinia to give advanced warning of Allied air raids. Numerous short-range radars were deployed to cover German bases in Italy.[42] The major problem for the *Luftwaffe* Signals Corps was a shortage of trained personnel in the theater. The *Luftwaffe* had lost its most experienced radar operators and air controllers in the Tunisia collapse and such men were hard to replace.[43] As the *Luftwaffe* frantically prepared a defense of Italy, the Allies clearly telegraphed their intentions to invade Sicily when they seized the island of Pantelleria in early June in a major amphibious operation supported by a massive two-week air bombardment. Pantelleria, an island south of Sicily, had a major airdrome and a radar station. Taking the island was vital to the Allies because the radar station there provided the *Luftwaffe* with early warning of Allied air strikes; and the airfield, which was within fighter range of Sicily's beaches, would be vital in supporting any attack on Sicily.[44]

An elaborate British deception operation to plant false plans on a dead man with the fictional identity of a British Marine major that indicated an Allied attack in the Balkans rather than on Sicily was put into action. The dead "courier officer" had supposedly died in an air accident off the coast of Spain, and the British made sure the body would wash ashore where it would be found by Franco's police, who would be sure to turn over the forged false documents to the local German agent.[45] This ploy got some attention from the Abwehr in Berlin, but was completely disregarded by the intelligence staffs in the theater.[46] The few air reconnaissance missions that the *Luftwaffe* was able to fly confirmed that there was relatively little Allied air and port activity in the eastern Mediterranean while the ports in Western North Africa were crammed with shipping. This indicated an attack was coming soon on either Sicily or Sardinia—possibly both. In the first week of July, Kesselring and von Richthofen judged that an attack on Sicily was imminent, German and Italian forces there were put on the highest state of alert, and continual reconnaissance was flown over the sea to try to spot the approaching Allied invasion convoy.

With a fairly clear knowledge of the British and American air forces in the theater provided by his intelligence staff, Von Richthofen and his senior commanders spent the weeks before the invasion of Sicily developing tactics to deal with the anticipated campaign. Von Richthofen believed that the only way the *Luftwaffe* could effectively oppose the Allied landing would be to attack the Allied shipping. If the *Luftwaffe* sank enough ships, the buildup of Allied forces on the beachhead might be slowed or stopped. In any case, attacking the most vulnerable part of the Allied logistics might induce a delay or stalemate in the ground battle.[47] With a numerically inferior force, the *Luftwaffe* developed tactics to spoof the expected Allied radar and fighter cover over the landing beaches and slip bombers and fighter bombers in to attack the ships. German bombers trained to first fly out over the ocean at altitude, drop to sea level where Allied radar could not pick them up, and change course. They would climb to an attack altitude of 2,000–4,000 feet just before initiating the final bomb run.[48] Due to the overwhelming Allied air superiority, the *Luftwaffe* planned to make most of its attacks at night to take advantage of the weak Allied night fighter force in the theater. In day attacks, fighters and fighter-bombers were expected to try to slip past Allied air patrols to bomb the ships. In defiance of Goering's orders to fight more aggressively, Von Richthofen ordered his aircraft to avoid contact with British and American fighter units. He hoped for 5 to 10 percent aircraft loss rate to the Allied fleet's guns and air defenses since any major air-to-air battles would soon destroy his force.

On the morning of 9 July, an Italian patrol plane spotted an Allied invasion convoy south of Pantelleria. The *Luftwaffe* was immediately informed and soon German and Italian reconnaissance aircraft were shadowing the Allied movements. The Italian and German forces on Sicily were placed on invasion alert and before dark the first air attacks on the convoys began. The attacks resumed as the Allies made eight landings along the Sicilian coast at dawn on the 10th. It was a "target rich" environment for the *Luftwaffe;* and in the first days of the invasion the bomber, Stuka, and fighter units using the careful tactics directed by von Richthofen managed to slip by the Allied air cover and sink a few destroyers and merchant ships in day and night attacks. German losses, mostly to anti-aircraft, were within the 5–10 percent loss rate that von Richthofen deemed acceptable attrition.[49] Given the huge scale of the Allied invasion, 1,365 warships, transports, and supply ships, the Axis air attacks which sank 12 ships during the landings had no significant impact on Allied operations. On the German side, however, it appeared very differently. From the first attacks on the Allied fleet Italian and German pilots reported a tremendous rate of success.

Battle Damage Assessment, or BDA in the current jargon, was the weakest part of German intelligence and analysis. Apparently the *Luftwaffe* simply accepted every pilot's claim when it came to strikes on shipping. In the confusion of battle, every near miss became a hit, slight damage was rated as severe, destroyers were mistaken for cruisers, and so on. Getting accurate intelligence during a battle is difficult enough, but since most of the German attacks were at night, anything but a wild guess was impossible. The 2nd Air Fleet reported that it had sunk 100,000 tons of Allied shipping on the first day of the Sicily landing.[50] By the end of the second day, von Richthofen recorded that the Italian and German air forces claimed 350,000 tons of Allied shipping sunk or disabled.[51] Because of the wildly optimistic reports, von Richthofen kept up the bomber attacks on the fleet for four days until increasing losses and stronger Allied air cover from newly captured Sicilian airfields compelled the *Luftwaffe* to give up. The *Luftwaffe* claimed to sink 500,000 tons of Allied merchant shipping and several warships in the Mediterranean in July 1943. The real figure was fourteen merchant ships (80,000 tons) lost and two destroyers.

Sometimes the *Luftwaffe*'s intelligence enabled the 2nd Air Fleet to score some tactical successes in the Mediterranean. As before the Sicily landings, *Luftflotte 2* figured that the Bay of Salerno would be a very likely place for the next Allied landing. A Salerno landing would give the Allies a major port on the Italian mainland, Naples, and it was within range of the Sicilian airfields, now in Allied hands. Thus, the Allies could be assured of land-based air superiority. The Germans stepped up their over-the-ocean air reconnaissance and, as with the Sicily landing, spotted the Allied invasion fleet at sea and headed for Salerno the evening before the landing. As in Sicily, the Germans were ready for the Allied landing and were able to attack with ground and air forces as soon as the Allies landed. The Germans, using the new precision-guided bombs (Fritz X and Henschel 293), were able to score some successes in the week after the landing to include one battleship disabled (HMS *Warspite*), two cruisers badly damaged, and three other vessels sunk. As before, the *Luftwaffe* grossly overestimated the damage inflicted on the Allies by air attack.[52] Despite full warning of the Allied attack, the Allied aerial superiority was so overwhelming that the *Luftwaffe* was in no position to seriously threaten the Allied landings.

After the Allies were on the Italian mainland, Von Richthofen sought the opportunity to use the small and ever-dwindling forces at his disposal to strike a major blow against the Allies, and *Luftflotte 2*'s intelligence staff provided him with an excellent opportunity. With only a handful of reconnaissance aircraft, *Luftflotte 2*'s coverage of Allied shipping was limited and sporadic. However, the *Luftwaffe* soon found a very vulnerable spot where maximum damage could be inflicted on Allied logistics—the port of Bari in southeastern Italy. As the Allies began their push up the Italian mainland, Bari became a major base area and depot. It served the needs of the British 8th Army and was the primary receiving port for the bombs, parts, gasoline, and other supplies needed for the American strategic bomber force (the 15th Air Force) being set up in nearby Foggia. The Allies maintained their air superiority, but the Germans soon learned that Bari was poorly defended. There were no Allied night fighters in the area and the port had just a handful of anti-aircraft guns. So confident were the British and Americans in their superiority over the *Luftwaffe* that the port was fully lighted in order to facilitate round-the-clock operations. Moreover, the port was packed tight with shipping. Whole convoys would have to wait for days before being unloaded at a port almost overwhelmed by the Allied shipping it was now expected to handle. The port of Bari became a focal point for German aerial reconnaissance with almost daily flights monitoring port activity through November.[53] The *Luftwaffe* made quick photo runs to Bari with some Me 210s that had recently been sent to the theater. Designed as a heavy fighter, the Me 210 was generally unsuccessful in that role. However, it was fast, and when the *Luftwaffe* could get the planes to work, they were reasonably successful in slipping through the very weak Allied defenses and getting current information to the air fleet headquarters.

Luftflotte 2 assembled virtually all the bombers in the theater—105 planes—and just after a convoy had arrived and the port of Bari was packed with shipping, von Richthofen struck on the night of 2 December 1943. In a carefully planned attack, the German bombers came in at low level over the Adriatic. The Allied air defense radar at Bari was quickly blinded by the German use of window. There was no Allied fighter defense, and the only effective resistance came from a few alert anti-aircraft crews on the merchant ships. The Germans, attacking by the light of flares, systematically bombed the ships and docks for over an hour and lost only two aircraft. After the first attack wave, the Germans did not even need flares because their bombs set off a tanker whose flames illuminated the helpless merchant shipping. In short order, the Allies lost sixteen ships destroyed, eight others badly damaged, and hundreds of thousands of tons of cargo. The port was so heavily damaged that it did not resume full operation until February. The noted naval historian, Samuel Eliot Morison, called the *Luftwaffe*'s raid on Bari "the most destructive air attack since Pearl Harbor."[54] Severe damage to the major Allied logistics base on Italy's east coast inhibited British army operations for several weeks and was also a temporary setback to the buildup of the 15th Air Force at Foggia. The Bari raid is a model of good tactical and operational intelligence and effective planning. Von Richthofen hoped that he could strike another damaging blow to the Allied forces in Italy, but the opportunity never came again. By springtime, almost all the *Luftwaffe*'s forces in Italy were withdrawn and sent to France to await the expected Allied invasion of France.

The *Luftwaffe* Faces the U.S. Strategic Offensive, 1942–1944— Operational and Tactical Intelligence

The AAF began its strategic bombing campaign in a modest fashion in the summer of 1942 with small attacks on German installations in France. In the summer and fall of 1942, the main U.S. bomber target was the complex of massive submarine pens located along the French coast. *Luftflotte 3*'s well-trained fighter units, such as JG 2 and 26 (JG is Jagdgeschwader or fighter wing), took on the American formations and scored some successes. The German commanders then applied the most basic form of intelligence—careful personal observation and analysis of the U.S. threat. After every clash with the American bombers, the combat reports were circulated among the fighter unit commanders and the *Luftwaffe* staff. The defensive firepower of the American bomber formations was considered and American methods and tactics analyzed from every angle to include typical bombing altitude, actions over the target, and return-flight tactics. After only a few raids the commander of one of JG 2's fighter groups, Major Egon Mayer, concluded that a frontal attack was the best means of taking on the American B-17s and B-24s. On the AAF's 23 November raid on the submarine pens at St. Nazaire, JG 2's FW 190s used the new tactics to shoot down four B-17s.[55] Adolf Galland, newly promoted as a general of the fighter forces, also analyzed the American methods and approved of the frontal attack, but modified it so that the fighters began their attack at an altitude slightly above the bombers—the famous "12 o'clock high" position. By the end of 1942, this had become the *Luftwaffe*'s standard fighter tactic against American bombers.[56]

Luftwaffe Intelligence had several other means to build up an accurate picture of their enemy. Shot-down American planes were carefully examined and, in the case of some only slightly damaged aircraft, were salvaged and shipped back to Germany to the technical intelligence unit at Rechlin. There the German technicians were able to put some of the American fighters and bombers back into full flying order and *Luftwaffe* fighter commanders given the chance to fly the American planes to get a "feel" for their enemies' equipment. The *Luftwaffe* Technical Intelligence branch published some very thorough studies of American aircraft shot down in German-occupied territory; but the *Luftwaffe*'s line officers complained that, while the studies were very good, the Technical Intelligence branch work was so detailed that the work of publishing and disseminating the information became much too slow. The *Luftwaffe* fighter units and air corps and air fleet staffs wanted pretty good information *immediately* instead of perfect information later.[57] The problem was solved in a typical military fashion— through informal back channels. The fighter pilots wanted to know a few basic facts to help them attack the American planes with the best chance of success: How much armor was on the plane and where was it positioned? Where were the fuel tanks located? What were the areas least protected by the defensive armament? Liaison officers from the *Luftwaffe* staff and from the air corps managed to pry the information out of the technical intelligence specialists and quickly pass it along to the fighter units. If a German fighter pilot was a good shot, he knew where to hit a B-17 or B-24 to bring it down.[58]

Intelligence Branch West, responsible for analyzing the AAF and RAF, published a steady stream of pamphlets, bulletins, and studies about all aspects of the enemy air forces. New radar and navigation equipment was analyzed and described as well as enemy aircraft capabilities and tactics. Bulletin Number 25 of the *Luftwaffe* Intelligence Staff–West, dated 23 February 1944, described the U.S. long-range escort tactics.[59] *Luftwaffe* Intelligence determined that the P-51 B Mustang could operate over Berlin or Munich at full power for 10 minutes and on cruise power for 46 minutes. Me 109 pilots were advised not to break contact with a P-51 by diving as the Mustang could easily outdive the German fighter. The pamphlet remarked: "in contrast to the P-38, the Mustang is easy to control in a high speed dive."[60] As American escort fighters penetrated deeper and deeper into the Reich, the *Luftwaffe* Technical Intelligence staff published pamphlets describing the capabilities of the most common Allied escort fighters to include the P-51, the P-38H, the P-47, and the Spitfire IX. Information on the weaponry, ceiling, fuel load, external tank capacity, and ranges with and without drop tanks of the main Allied fighters was circulated to all the *Luftwaffe* units in the Reich and in the West.[61]

The *Luftwaffe* worked hard to develop a comprehensive understanding of the American Army Air Forces' order of battle, with information on unit strength, location, and training levels. One of the most productive and accurate sources of intelligence was Dulag Luft, the *Luftwaffe*'s prisoner interrogation center. From a modest start in 1939 with a handful of prisoners held at a farmhouse, Dulag Luft processed 3,000 Allied airmen in 1942, 8,000 in 1943, and 29,000 in 1944.[62] By 1944, Dulag Luft employed 60–65 *Luftwaffe* interrogators, men highly fluent in English and experts in fields such as fighter or bomber units. Incoming prisoners were first studied by psychologists who determined the most productive approach to break them down. Some were handled roughly and kept in strict solitary confinement. Others were handled in a friendly manner. The German master interrogators, some of whom were legendary for their ability to pry information from captured airmen, had no need of torture or Gestapo methods to obtain a vast amount of helpful data. Armed with bits of information about a unit—such as the commander's nickname or the recent unit movement —friendly German soldiers confronted often naive airmen with complete knowledge about them and their organization. Over a friendly chat, an American airman would let drop important details about his training or unit.[63] The Germans were helped enormously by the tendency of U.S. pilots to carry letters, diaries, and photos of their squadron, operations orders, and so on with them—against AAF orders. Such documents recovered from plane wrecks and prisoners were extremely helpful.

Each interview was carefully compared against others, and all the bits of information soon added up to a very complete and accurate picture of the American bomber and fighter units flying over German territory. In fairly short order, the *Luftwaffe* developed a sound understanding of AAF training, unit organization, and order of battle. One senior *Luftwaffe* officer commented that the American aircrews, in particular, were naive under questioning.[64] All the *Luftwaffe* interrogators needed was to get American pilots talking. Apparently many Americans, fighter pilots in particular, simply could not keep quiet about themselves, their unit, and their flying exploits and inadvertently dropped details useful in analyzing *Luftwaffe* flak and fighter tactics. All too often, pilots were willing to tell how they were shot down and by what weapons.[65] In 1943 and 1944, a hundred interrogation reports a day on the RAF and AAF were flowing out of Dulag Luft to the *Luftwaffe* staff. Combined with the excellent work done by General Wolfgang Martini's specialized signal intelligence units that monitored RAF and AAF radio traffic, the *Luftwaffe* fought the Americans abetted with an accurate picture of USAAF organization, training, and tactics.[66]

The star performer of the *Luftwaffe*'s intelligence organization was the *Luftwaffe* Signal Corps, commanded by General Wolfgang Martini from the start of the war to the end. Martini had tremendous technical talent and outstanding leadership ability. As the commander of Signal Troops, he was responsible for the *Luftwaffe*'s radios, radars, electronic warfare systems, and signals intelligence. The story of the *Luftwaffe*'s signal forces and electronic warfare program has been told in detail in several fine works—most notably Alfred Price's *Instruments of Darkness*—and need not be told in detail here. However, for the most part, the *Luftwaffe*'s signal units were an important source for accurate intelligence.[67] The first warning that the *Luftwaffe* would receive of any American bomber raid came from the *Luftwaffe* signals intercept detachments stationed in France and Germany. From the volume and type of radio traffic the *Luftwaffe* would know an hour or two before a major mission started. Since the Germans also knew that American bombers took 1½ to 2 hours to organize their large formations after takeoff, the *Luftwaffe* had 2 to 3 hours of warning before it could expect U.S. bombers to fly into German-held airspace. The important role of *Luftwaffe* signals intelligence is illustrated by the disastrous experience of the AAF raid on Ploesti on 1 August 1943. This raid was to strike a crippling blow to Germany's largest single source of oil—the Ploesti oilfields in Romania. On the morning of 1 August 1943, 178 B-24s took of from Benghazi in North Africa to strike Ploesti. A *Luftwaffe* signals intercept station in Greece intercepted the U.S. radio traffic at takeoff and immediately alerted all the flak and fighter units in the Balkans that a large U.S. raid was coming from the south. German flak gunners and fighters were fully prepared for the United States attack and 53 planes, 30 percent of the total force, were shot down. Damage to the refinery was moderate and, thanks to the deadly *Luftwaffe* defense, Germany's most vital installation continued to operate.[68]

Luftwaffe signal troops came to know American radio procedure quite well and were able to pick up considerable intelligence on the AAF simply by monitoring the unit radio chatter. A summary of *Luftwaffe* signals intercept intelligence on the RAF and AAF from March and April 1944 establishes a very accurate order of battle for the 8th Air Force Bomber Command and associated bomber groups with specific air bases and air divisions. Experienced German operators came to know the peculiarities of American and British radio procedure, quickly identified call signs and monitored every change in call sign, radio frequency, and procedure. The signal intercept units could identify units in transit and training and could make some pretty sound judgments on the standard of training in specific units from their radio discipline and procedure.[69] Combining this information with interrogation reports from Dulag Luft, the *Luftwaffe* was able to maintain an accurate picture of AAF strength, order of battle, and capabilities during the course of the American strategic bombing offensive.

For the tactical air battle in the skies of Western Europe, the radar operators and air controllers of the *Luftwaffe* signal branch played the key role in the air defense of the Reich. By 1942, when the Americans began attacking targets in occupied France, the Germans had already established an integrated air defense system in the coastal zones of occupied Europe. As the British mounted ever larger night-bombing raids against Germany in 1942 and 1943, the *Luftwaffe* reorganized the Reich air defenses and built up the homeland defense forces with additional fighter and flak units. By the time the Americans began flying deep missions into Germany in 1943, they faced a well-integrated and coordinated air defense system and numerous radar stations able to continuously track raids flying from England. As in the Battle of Britain, the well-trained *Luftwaffe* ground controller became one of the most important battle assets as he directed fighter units to intercept the Americans and kept the fighter units continually informed of the U.S. bombers' altitude and course. The *Luftwaffe* ground controllers and fighter pilots were highly effective in directing the defense during the disastrous U.S. raids on Schweinfurt and Regensburg in August 1943 when the 8th Air Force lost 60 of 376 bombers it sent out. After further heavy losses in October, the 8th Air Force called off deep unescorted bomber raids into Germany for several weeks.[70]

When the Americans resumed the bomber offensive deep into Germany in January 1944, this time protected by long-range escorts, German fighter defenses became even more dependent on the radar operators and ground controllers. The *Luftwaffe* tried a variety of tactics to get past the escorts and attack the bombers, but most of them failed. For example, the German defenders would first send in a squadron of Me 109s to attack from the side and rear of the formation and hopefully draw the escorts away from the bombers so a group of Fw 190s could make their characteristic head-on pass and dive away. This rarely worked as American fighter units refused to take the bait and go after the Germans. When American escorts did pursue the decoy groups, only some planes would be detached from the formation—which always had a couple of squadrons keeping close escort. The Americans soon introduced fighter sweep tactics and flew escort well in front of the bomber units and were able to catch German fighters often in the process of forming up. By the spring of 1944, the only effective fighter tactic against the American bombers was to wait for a break in the escort coverage and bore in on the unescorted bombers. The American fighters had limited cruising ranges and could not accompany the bombers all the way to the target and back while throttled back to match the bombers' cruising speed. The only efficient means to escort the bombers was to have the shorter ranged fighters, such as the Spitfires and P-47s, cover the bombers for the first leg of the mission, and somewhere around the German border turn them over to the longer ranged P-51s that would take them to the target. On the trip back, the bombers might again be met by the shorter ranged fighters to provide cover for the last leg. It required complicated planning and coordination to make the fighter-escort system work and there were many instances when weather or navigation error would cause an escorting fighter group to miss its rendezvous with the bombers. That was the point at which the U.S. bombers were most vulnerable. An experienced radar operator and ground controller could identify a failed rendezvous and direct the German fighters in for easier kills. While the escort system generally worked, there were a few times when it failed.

Because of its corps of capable controllers, the *Luftwaffe* fighters were able to take bloody advantage of mistakes and inflict heavy losses on bomber groups. However, accurate intelligence, first-rate air controllers, and fine tactics cannot win the battle if the means—namely good pilots and superior aircraft—are not available. To win the fight against the Americans, the Reich air defense force needed experienced fighter commanders and skilled pilots—and these were in very short supply by the spring of 1944 when the American bombing offensive got fully underway thanks to the strategic miscalculations made by Hitler, Goering, and the *Luftwaffe* staff early in the war. In every aspect of operations, the Allies had the advantage in numbers, equipment, and aircrew training; and the situation was basically hopeless. Indeed, Germany's minimally trained fighter pilots not only found it difficult to hold formation and fly precise attacks against the Americans, they could barely navigate or land their aircraft safely. A huge part of the *Luftwaffe* pilots' losses in 1944 came not from combat but simply from operational accidents. As the year progressed, the fighter forces of the *Luftwaffe* became less and less capable.

The *Luftwaffe* regained one important intelligence asset late in the war. During 1943 and most of 1944, the margin of Allied air superiority was so great that the *Luftwaffe* was rarely able to slip a long-range reconnaissance aircraft past Allied air defenses. In the vital weeks before D-Day, *Luftwaffe* strategic reconnaissance was blind to the buildup of Allied invasion shipping in British ports. During the 1944 battle for France, the *Luftwaffe* could not report on Allied movements or airfield activity. This changed in September when the *Luftwaffe* deployed the first of several Arado 234 jet reconnaissance planes, a version of the just-developed jet bomber.

With a speed of over 460 mph and a regular operating ceiling of 29,000–39,000 feet, the Arado could fly photoreconnaissance missions even over Britain in relative safety from the piston-engined Allied fighters.[71] With Arados based in northern Germany, the *Luftwaffe* was able to gather accurate intelligence on the American and British tactical air forces stationed at airfields in Belgium and northern France in anticipation of a great blow against the RAF and AAF.

Armed with accurate knowledge of Allied airfields, on 1 January 1945, over 1,000 German fighters—virtually the entire fighter reserve of the *Luftwaffe*—made a surprise dawn attack upon dozens of airfields across Belgium with the hope of inflicting a crippling blow against the Allies. Despite good intelligence and careful planning, things went very wrong for "Operation Bodenplatte." Some German units managed to successfully shoot up British and American airfields, destroying dozens of fighters and medium bombers on the ground. Other groups got lost and missed the target, and the inexperienced German fighter pilots became easy prey for the Allied fighters on patrol. Some *Luftwaffe* units attacked the Allied airfields but were shot to pieces by their own flak as they reentered German airspace—an example where coordination had clearly fallen through. In addition, Allied flak was fierce and claimed dozens more Germans. At the end of the day, the Allies had lost about 300 planes on the ground— aircraft that were easily replaced from reserves within a couple of weeks. On the other hand, the *Luftwaffe* lost more than 300 planes and pilots, including some of the last experienced group and squadron commanders. It was a disaster for the *Luftwaffe* and signified the end of effective operations for the Reich fighter force—another example of how good intelligence cannot deliver a victory without the combat means.[72]

Conclusion

In summary, the *Luftwaffe*'s intelligence on the U.S. Army Air Forces reflects the general performance of the German military as a whole. At the strategic level, the Germans crippled themselves. At the operational and tactical levels, the Germans could be very good. Of course, the failure of the *Luftwaffe* leadership and Hitler's inner circle to understand the potential of American air power early in the war was not due to any lack of good intelligence or analysis—as noted, there was plenty of hard data and good analysis presented by many experts who tried to warn Hitler and Goering. It was characteristic of Hitler, Goering, and senior officers such as the *Luftwaffe* Chief of Staff General Hans Jeschonnek to simply ignore intelligence that conflicted with their deeply held notions that a decadent and democratic America simply could not produce vast numbers of high-quality aircraft and pilots. All German strategic planning until 1942 was based on the dogma that the war would be short and victorious, and that a massive expansion of the aircraft industry and pilot-training program would not be necessary. By the time the Nazi leadership understood their strategic position, it was too late to catch up. The uncoordinated nature of the German intelligence organization did not help matters much. Accurate intelligence on the Americans from the military attaché was cancelled out by reports from other intelligence agencies that downgraded American potential and supported Hitler's worldview.

It was our good fortune that German strategic planning was so badly informed—a mistake that helped win the war for the Allies. However, Americans ought not to be smug about the Nazi failure to understand strategic intelligence. A massive and disjointed intelligence bureaucracy with agencies often working at cross-purposes is not a unique characteristic of the Nazi state, it also flourishes in great democracies. One can look at the state of American strategic intelligence concerning the terrorist threat on 11 September 2001. The CIA, FBI, Immigration and Customs Services, and many other agencies all collected intelligence on radical Islamic terrorists who had been attacking Americans for a decade. There was plenty of information but no effective coordination and no single agency responsible for analyzing the threat. Indeed, under the stovepiped nature of our intelligence system, the FBI and CIA were not allowed to share vital information. The result was one of the most dramatic intelligence failures of modern times. Whether we have reformed our intelligence system and corrected major flaws is still an open question.

Ignoring or downplaying intelligence that conflicted with the prejudices of the leadership was a common trait of Nazi strategic decision-making. Again, this is a trait not confined to the Nazis or to totalitarian societies. On several occasions in the last century, American military and civilian leaders have ignored and downplayed vital strategic intelligence with disastrous consequences. We should remember autumn 1950 when strategic intelligence about a massive Chinese intervention in Korea came in the form of captured Chinese soldiers—and was disregarded by the senior military commanders and the CIA who held fast to their conviction that Chinese intervention was unlikely. That mistake almost caused the destruction of an American army. Certainly the prime example of American strategic hubris is the manner in which our best and brightest military and political leaders jumped into the Vietnam War without a serious attempt to analyze the North Vietnamese leadership, strategy and will. American military leaders saw no relevance in the recent war the French had fought and began bombing North Vietnam with a plan based on simplistic theories held as dogma since the 1930s. Today, as the guerrilla war in Iraq continues five months after the end of major combat operations, it seems that our prewar assessments about the will of Saddam Hussein's loyalists to resist the U.S. occupation as well as our predictions about rebuilding the Iraqi economy and infrastructure were incredibly optimistic.

In the case of tactical and operational intelligence, the *Luftwaffe* performed very creditably. Indeed, some senior *Luftwaffe* commanders proved quite adept at using intelligence to gain the combat advantage. The theater and area commanders often had a much better grasp of Allied capabilities and forces than their superiors in Berlin. Unlike the OKW staff, Kesselring and von Richthofen were not taken in by Allied deception operations. The Sicily and Salerno landings were expected and German defenders given ample warning. As we move today to an ever greater centralization of command authority and a preference for micromanagement of military operations, we might do well to remember that the operational commander close to the front might have a much better appreciation for the theater situation than the higher command levels.

Air fleet commanders proved they could use their intelligence assets to great advantage. The preparation and execution of the 2nd Air Fleet's raid on Bari, which was a serious blow to the Allies, is a model of the proper use of operational intelligence. In northern Europe and the Mediterranean many thousands of British and American airmen died because General Martini's efficient Signals Corps gave advance warning to German defenders at places like Ploesti. In the battle for the airspace of the Reich in 1943–44, German fighter pilots often inflicted heavy losses on American bomber groups because the *Luftwaffe* radar operators and air controllers were very good at their jobs and had good equipment. The ground controllers understood American tactics well enough to direct the fighters into the attack at the right place and time. As long as the Germans had adequate trained forces to fight the battle, the various branches of *Luftwaffe* intelligence were an effective force multiplier. Eventually, however, the Germans simply ran out of pilots—which made even the best intelligence irrelevant.

Some of the problem areas of *Luftwaffe* intelligence, such as the lack of accurate BDA, are still very much with us today. Pilot reports of damage inflicted upon the enemy from every conflict and air force, from Korea to Vietnam to the Falklands to the 1999 air campaign over Kosovo, have usually been wildly inaccurate. On-the-ground postwar assessments generally deflate pilot claims. Of course, in the stress and excitement of aerial combat, it seems natural for pilots to exaggerate. One suspects that pilot egos and culture tend to cross national and temporal boundaries.

Another consistent problem with intelligence analysis, one that applied equally to the Allies and the *Luftwaffe,* was in assessing the enemy's aircraft production and operational strength. Through much of the war, the Germans used complicated economic models that put U.S. and British production far below their actual levels. The *Luftwaffe* was much better at judging the operational strength of the AAF and RAF because of good order-of-battle information, but even then the *Luftwaffe* Intelligence Branch tended to overestimate the U.S. bomber strength by a factor of about 30 percent.[73] The RAF and AAF intelligence branches, employing similar economic models, were 30–80 percent off on their estimates of German aircraft production throughout most of the war.[74] The moral of the story is that accurate economic modeling is extremely hard, and even a lot of very smart analysts armed with a lot of information can still be way off in their conclusions.

No matter how much we hear today about the instantaneous communications, "the digital battlefield," "network centric warfare," space surveillance that can read a license plate at 100 miles, UAVs that can loiter over the battlefield for days, and the whole array of gadgets at our disposal to collect intelligence—intelligence collection and analysis is still very much an art, not a science. The Germans put themselves in an impossible situation by ignoring strategic intelligence in the first half of the war. Consequently, by 1944, the U.S. air forces in the Mediterranean and European theaters enjoyed enormous superiority over the *Luftwaffe* in terms of aircraft numbers and aircrew training. The Allies also possessed a tremendous intelligence advantage in Ultra. In the end, the *Luftwaffe* and the German war machine was overwhelmed by superior forces used with great skill. Yet, at the operational and tactical levels, from 1943 through 1944, a rapidly declining *Luftwaffe* was able to use its knowledge of American forces and tactics fairly effectively to inflict heavy losses on the AAF.

Notes

1. James S. Corum, *The Luftwaffe: Creating the Operational Air War 1918–1940* (Lawrence: University Press of Kansas, 1997), 58–72.

2. James S. Corum, *The Roots of Blitzkrieg* (Lawrence; University Press of Kansas, 1992), 158.

3. The major U.S. Army journals of the interwar period, such as the *Infantry Journal, Field Artillery Journal, and Cavalry Journal* all devoted considerable space in covering the German operations of the Great War and current German military thought. German articles were often translated and German books reviewed. The whole tone is one of great respect for the German Army and its professional tradition.

4. Corum, *Roots of Blitzkrieg,* 82–83, 90–91.

5. Corum, *Luftwaffe,* 98–101.

6. Major Helmut Wilberg, Reisebericht 1925–26, doc. RH 2/1820, 199, Bundesarchiv/ Militärarchiv Freiburg (BA/MA).

7. Major von dem Hagen, Reisebericht, November 1928, doc. RH 2/1822, ibid.

8. Captain Speidel, Berichte, ibid.

9. Corum, *Roots of Blitzkrieg,* 157–58.

10. For an excellent overview of the German intelligence system, see David Kahn, *Hitler's Spies: German Military Intelligence in World War II* (New York: MacMillan, 1978).

11. On General von Boetticher, see Alfred Beck., *Hitler's Ambivalent Attaché* (Washington, D.C.: Potomac, 2005).

12. Ibid., 81.

13. Von Boetticher to Reichswehrministerium, Anlage 2, Report 20/1933, 25 Oct. 1933, German Records, T-120 2741/5863, National Archives and Records Administration, Washington, D.C.

14. Von Boetticher to Reichswehrministerium, "The American Army in the Fiscal year 1937," Anlage 1, Report 27/1937, 25 July 1937, ibid.

15. Beck, 81–82.

16. Papers of D. A. Dickey, U.S. engineer for USSTAF, "Report on Survey of Propeller Information at Engineering Plant of Messerschmitt at Oberammergau," June 1945, USAF Historical Research Agency (USAF/HRA), Maxwell AFB, Ala.

17. On Peter Riedel, see Beck, 114–22.

18. Ibid., 152.

19. Ibid., 154–58.

20. Ibid., 182.

21. David Irving, *Goering: A Biography* (New York: Avon, 1989), 338.

22. Kahn, 84.

23. Siebel's letter to Goering is cited in Lt. Gen. A. D. Andreas Nielsen, *Die Nachrichtenbeschaffung und –auswertung fuer die deutsche Luftwaffenfuehrung,* Karlsruhe Study, 1957, doc. K 113.106.170, 184–88, USAF/HRA.

24. Ibid., 183–85, 197–98.

25. Horst Boog, *Die deutsche Luftwaffenfuehrung 1935–1945* (Stuttgart: Deutsche Verlags-Anstalt, 1982), 28–29.

26. "Organization of IC, GAF Intelligence Staff 1942," in "GAF Intelligence and its Chiefs 1938–1945," U.S. 8th AF Report, Fall 1945, doc. 533.6314-1, USAF/HRA.

27. Kahn, 210–11.

28. Horst Boog, "German Air Intelligence in World War II," *Aerospace Historian* (June 1906)1122.

29. Cited in Maj. Bradford Shwedo, *XIX Tactical Air Command and Ultra,* Cadre paper no. 10 (Maxwell AFB, Ala.: Air University Press, 2001), 17.

30. For a very negative assessment of Schmid, see Boog, *Die deutsche Luftwaffenfuehrung,* 81–82, 96–97.

31. On *Luftwaffe* intelligence early in the war, see Kahn, 382–84.

32. Boog, *Die deutsche Luftwaffenfuehrung,* 81; Kahn, 385.

33. Boog, *Die deutsche Luftwaffenfuehrung,* 81.

34. The documents—Directive no. 77, 24 Jan. 1945, and memo from Spaatz, 5 Jan, 1945, and commentary are found in Air Ministry, "Report on Investigation of OKL Intelligence," 506.6314-11, USAF/HRA.

35. Dr. David Mets, a biographer of Spaatz and author of *Master of Airpower: General Carl A. Spaatz* (Novato, Calif.: Presidio, 1988), has examined the forged documents and says they are about "70% correct."

36. General Paul Deichmann, ed. *Die deutsche Luftwaffe in Italien,* Karlsruhe Study, April 1956, doc. K 113.310.8, 1943–1945, pt. II, 7, USAF/HRA.

37. William Green, *Warplanes of the Third Reich* (New York: Galahad, 1970), 425–26.

38. Ibid.

39. Deichmann, pt. II, 1–3, and pt. III, 45–47.

40. Ibid.

41. Nielsen, 173–75.

42. Deichmann, pt. II, 37–40.

43. Ibid., 40–41.

44. Ibid.

45. Ewan Montagu, *The Man Who Never Was* (Philadelphia, Pa.: Lippincott, 1954).

46. Nielsen, 178–79.

47. Diary of Field Marshal von Richthofen, entry of 12 July 1943. Diary copied with permission of the von Richthofen family.

48. Deichmann, pt. II, 14–15, 53.

49. The Germans lost at least 27 aircraft on the first two days of the campaign. See Brig. Gen. C. J. Molony, *The Mediterraenian and Middle East,* vol. 5 (London: HMSO, 1973), 66.

50. Von Richthofen diary, 10 July 1943.

51. Ibid., 11–12 July 1943.

52. Ibid., 9 Sept. 1943.

53. For a record of *Luftwaffe* reconnaissance over Bari, see *Luftflotte 2* Intelligence reports in BA/MA doc. RL/II 304, Nov. Dec. 1943, and *Luftflotte 2* Lageberichte BA/MA/2/II/369 Nov. Dec. 1943.

54. Samuel Eliot Morison, *History of United States Naval Operations in World War II,* vol. 9 (Edison, N.J.: Castle, 1954), 319, 322.

55. John Weal, *Jagdgeschwader 2 "Richthofen"* (Oxford: Osprey, 2000), 91–93.

56. Ibid., 94.

57. "GAF Intelligence in World War II, 8th Air Force intelligence report," 16 Nov. 1945, doc. K 113.60, 7, USAF/HRA.

58. See Kahn, 157–59, for the *Luftwaffe*'s analysis of shot down B-17s.

59. "OB Luftwaffe, Fuehrungsstab Ic, Einzelnachrichten des Ic Dienstes West Nr. 25. Bemerkungen zum Feindeinsatz Nr. 6," 23 Feb. 1944, doc. K 113.60, 11–12, USAF/HRA.

60. Ibid., 14.

61. Techniches Amt, "Feindliche Begleitjaeger," 15 Mar. 1944, doc. K 113.60, USAF/HRA.

62. "The Luftwaffe Interrogators Dulag Luft-Oberursel." Website: <www.merkki.com/new-page-2.htm>, contains photos and details about the German interrogators.

63. German interrogation techniques are outlined in Kahn, 138–40.

64. Nielsen, 88.

65. A good example of a U.S. pilot interrogation report is found in Kahn, 140.

66. A detailed look at the whole prisoner interrogation process is found in Raymond Toliver, *The Interrogator: the Story of Hanns Scharff, the Luftwaffe's Master Interrogator* (Raymond Toliver, 1978).

67. Alfred Price, *Instruments of Darkness: The History of Electronic Warfare* (New York: Scribners' Sons, 1978).

68. Kahn, 211–12.

69. Captured G.A.F. Sigint Service document, No. 1 Company of Luftwaffe Wireless Listening Regiment West, 11 March–10th April 1944. Captured in France on 17 Sept. 1944, doc. 506.6314, USAF/HRA.

70. Martin Middlebrook, *The Schweinfurt-Regensburg Mission* (London: Penguin, 1983).

71. Green, 54–55.

72. For a good account of Operation Bodenplatte, see Danny Parker, *"To Win the Winter Sky": Air War Over the Ardennes 1944–1945* (Pennsylvania, Pa.: Combined Books, 1994), 406–49.

73. The *Luftwaffe* Intelligence Branch in February, March, and April estimated 8th Air Force bomber strength at 4,860–4,950. This was approximately 1,000 planes too many. See SHAEF Intelligence Party OKL, Intelligence Report No. 41, 13 Aug. 1945, doc. 506.6314A-41, USAF/HRA.

74. Boog, *Die deutsche Luftwaffenfuehrung,* 101. On the difficulty of doing accurate BDA analysis on the German aircraft industry, see John Kreis, ed., *Piercing the Fog: Intelligence and Army Air Forces Operations in World War II* (Bolling AFB, Washington, D.C.: Air Force History and Museums Program, 1996), 202–3.

The Aircraft That Decided World War II: Aeronautical Engineering and Grand Strategy, 1933–1945, the American Dimension

John F. Guilmartin, Jr.

The purpose of this essay is to connect, from an American perspective, two propositions: that air power was critical to the conduct and outcome of World War II, and that aircraft design contributed the crucial role in the process. There is nothing controversial about the first proposition. Historians and theoreticians may debate the decisiveness of strategic bombing, but few would deny the decisiveness of air power in the generic sense, if for no other reason because control of the air invariably provided an essential ingredient of victory in the battles and campaigns comprising World War II and whose cumulative effects of air power determined its flow and eventual outcome. Indeed, the war's only campaign of major strategic consequence won without benefit of air superiority was the U.S. Navy's submarine campaign against Japanese shipping, a fact that is in part testimony to Japanese weakness in the air.[1]

To underline air power's importance, consider the battles and campaigns principally responsible for shaping the course of the conflict: the Battle of France, the Battle of Britain, the early German victories on the Eastern Front; the Japanese Centrifugal Offensive; Midway; Guadalcanal; Stalingrad; El Alamein; the Tunisian campaign; the Battle of the Atlantic; D-Day; the Normandy campaign, breakout, and pursuit; the Destruction of Army Group Center; the New Guinea campaign; the Central Pacific campaign culminating in the Battle of the Philippine Sea and the seizure of the Marianas; the Battle of Leyte Gulf; and the bombing of Japan. In every case, victory was secured from the air, was dependent upon its control, or both.[2]

Extending our analysis to smaller engagements of strategic consequence yields a number of naval actions in which air power played a negligible role. Significantly, most were fought at night, testimony to the vulnerability of warships to daylight air attack even early in the war.[3] Notable among them were a series of night surface engagements between Japanese and American forces during the Guadalcanal campaign.[4] These engagements took place at night precisely because the Japanese Navy sought cover of darkness to negate U.S. air power. Note, too, that aerial reconnaissance (or the failure thereof) played a major role in most if not all of these engagements and that American superiority in the air was essential to Allied victory in the overall campaign.[5] The other exceptions are partial and qualified. The Royal Navy was able to fight convoys through to Malta and Murmansk under heavy air assault and with little or no air cover, but at great cost. Indeed, Japanese successes in the Guadalcanal campaign aside, the only naval victory of consequence won by surface forces unaided by air and fought within range of enemy airfields was the 26 December

1943 Battle of North Cape in which the Royal Navy sank the German battle cruiser *Scharnhorst:* confined to its Norwegian airfields by abominable weather, the *Luftwaffe* failed to intervene.[6]

We should also note that a number of the war's most important campaigns were fought entirely in the air, notably the Battle of Britain, the Combined Bomber Offensive, and the strategic bombing campaign against Japan. I would argue that all three were strategically decisive, an assertion that raises the question of strategic bombing's effectiveness in World War II. I will address that question later, but first let me make a fundamental point: that war in the air is inherently different from other forms of warfare and that we do not truly understand it, even today, over a half-century after V-J Day. A key problem is that we approach strategic bombing with the implicit assumption that air campaigns and battles can be judged using the vocabulary, criteria for success or failure, and analytical framework used to evaluate warfare on land and at sea. I contend that the appropriate criteria for judging strategic air campaigns, at least, are quite different and that in consequence the results of the debate so far are of dubious validity.

As evidence of our incomplete understanding of the nature of aerial warfare, consider the general lack of consensus—or even awareness—of what constitutes an air campaign. To illustrate the point, consider the last major Axis campaign victory of World War II. When asked to name the campaign in question, most draw a momentary blank and then think of the Battle of the Bulge before recalling that it ended in German defeat. It was, in fact, the Battle of Berlin, the effort by Royal Air Force (RAF) Bomber Command between November 1943 and March 1944 to destroy Berlin, repeating Hamburg's destruction the previous summer.[7] In the process, the British inflicted considerable damage on Germany to be sure, but the result was unequivocally a German victory: Bomber Command called off its offensive after an accumulative loss of nearly 1,100 aircraft, almost all of them four-engine bombers.[8] Indeed, the final battle of the campaign, the 30 March Nuremberg raid, was one of the largest air battles of the war, if not *the* largest,[9] and a signal German victory. There is no denying the strategic importance of the British defeat in terms of expenditure of resources and in lives lost, a cost made all the more painful by the fact that the lives in question were those of a highly trained and strategically important elite aircrews, yet it was not a typical battle or campaign.

To expand on the point, consider the nature of the Combined Bomber Offensive. We ordinarily think of it as a campaign, but it was in fact something larger, for it contained within it operations that clearly qualify as campaigns in their own right: RAF Bomber Command's area bombing of German cities; the United States Army Air Forces' Unescorted Daylight Strategic Bombardment Campaign of 1943; the 1944 campaign against German sources of oil and fuel production; and Big Week, the U.S. Army Air Force (USAAF) effort in February 1944 that forced the *Luftwaffe* fighter force to accept battle and, ultimately, defeat. The lesson is evident. Not only does the terminology that we have inherited from land and naval warfare fit war in the air poorly, it carries with it analytical baggage that distorts analysis.

To further underline the inherent difference between war in the air and war on the surface, one can argue—and I do—that World War II in the air comprised a unitary global conflict in ways that the war on land and at sea did not. On land and at sea, the war can be usefully divided into theaters and fronts: the European Theater; the Eastern Front; and the Mediterranean, China-Burma-India, Southwest Pacific, and Central Pacific theaters, and so on. By and large, there was little movement of ground forces from one to another. With the partial exception of the Germans, who used interior lines to transfer their strategic reserves, once ground forces were committed to a theater or front they stayed there. The same general point applies to naval forces to an only slightly lesser degree. But what about air forces? Air forces *were* transferred from theater to theater with some frequency. The USAAF transferred much of its deployed force structure from Britain to North Africa in the autumn of 1942. The Germans shifted air units from front to front far more frequently than their ground reserves. The Japanese Army Air Force transferred much of its strength from the Home Islands and Manchuria to the Southwest Pacific in 1942–43. The manner in which air reserves were deployed in certain critical instances leads me to the conclusion that at least senior Allied leaders implicitly understood that the air war was indivisible by theater. Let me make the point by example.

For the U.S. Navy, the series of actions in May and June of 1942 that culminated in the Battle of Midway were the most critical of the war, a fact of which senior commanders were keenly aware well before the fact. As they were also keenly aware, fleet carriers were *the* critical operational asset. At the beginning of May 1942, the U.S. Navy had five fleet carriers capable of flight operations: *Lexington, Enterprise, Hornet, Yorktown,* and *Wasp.*[10] Of these, *Lexington* was sunk at the Coral Sea and *Enterprise, Hornet,* and *Yorktown* fought at Midway. Where was *Wasp*? More precisely, why was *Wasp* not at Midway? Because it was delivering Spitfires to Malta! Those responsible for sending *Wasp* to the Mediterranean clearly understood the global nature of air power, and it is worth noting that American forces began to establish ascendancy in the air in the Pacific shortly thereafter, at the same time the British were taking the *Luftwaffe*'s measure in North Africa and the Mediterranean.

Returning to the definitional problem, the argument that strategic bombing failed in World War II is generally made by evaluating the results of individual campaigns in isolation. Most often cited are the USAAF efforts in the summer and autumn of 1943 to collapse the German war economy by unescorted daylight bombardment and RAF Bomber Command's night area bombing of German cities. While it is true that neither succeeded in achieving its stated objectives, both forced major reallocations of German resources that might have been more profitably used elsewhere. Of far greater importance to the subsequent course of the war, both campaigns, and in particular the American effort, depleted vital German resources that could not be replaced within available time constraints, most critically skilled fighter pilots. Thus while they may not have been victories within the analytical framework borrowed from warfare on land and at sea, both had long-term consequences that contributed powerfully not only to allied victory in the air, but also to the defeat of Nazi Germany.

So with an understanding that there is much we do not understand about the nature of air warfare, let us turn to my second proposition, that aircraft design was a key variable in determining the strategic effect of air power. This proposition, like my first one, is uncontroversial, although here the lack of controversy is mostly due to a lack of systematic examination of the problem. Almost by definition, well-designed aircraft have superior performance and should thus bestow to their possessors tactical, and therefore strategic, advantages, or so logic would dictate. But there is a danger in this assumption, for it is easy to conclude that strategic advantage obtained in the air must have flowed from superior design and that is not always the case.

In fact, the seemingly straightforward relationship between quality of design and tactical advantage on the one hand and strategic effect on the other turns out to be anything but. As a multitude of cases demonstrate superiority in numbers or employment tactics, acting together or independently, can do much to offset performance disadvantages. That part of the puzzle is generally understood. Not so well understood or systematically explored is the fact that design determines much more than performance in the narrow sense: speed, maneuverability, range, offensive capabilities, resistance to battle damage, and the other factors that influence tactical effectiveness. By predicting cost and ease of production, design sets limits on how many of a given design can be built with the fiscal and human resources available. In controlling reliability and ease of maintenance, design has a major influence on in-commission rates. In establishing handling characteristics, in simple terms how easy or difficult an aircraft is to fly, design exercises a powerful influence on operational wastage. Finally, the design must be suited for the particular circumstances under which the aircraft is to be employed, and here a single performance parameter may be critical. To cite an obvious example, a bomber which is a superior design in every other respect, but which lacks the range to reach its targets is strategically useless.

We are not helped much in our inquiry by the secondary literature, for little attention has been paid in detail to the connection between aeronautical design, tactical operation, and strategic impact. A great deal has been written about the impact of air power on World War II by theater, campaign, and battle, but few general accounts pay much attention to aircraft performance, let alone design. Similarly, much has been written about the aircraft with which the war was fought, their design histories, what they were like to fly, and how successful they were in combat, much of it for a buff audience. But while capturing an enormous amount of valuable information, this literature rarely addresses strategic issues. As a result of this divide in the literature, attempts to relate aircraft design to strategic effect are rare and generally limited to a single campaign or battle. The Battle of Britain is well served in this regard, but is very much the exception to the rule.[11]

The intersection between aircraft design and strategic effect is an enormous topic, and in addressing it I confronted major problems. The key question was which aircraft to analyze, and it struck me that it might be useful to begin by ranking World

War II aircraft according to their strategic importance. Such a rank-ordering would not only reduce the scope of the inquiry to manageable proportions, it would, or so I hoped, provide an analytical lens through which to selectively identify and evaluate those performance characteristics that were strategically most important. Having identified the critical performance parameters, I could then examine the design processes that produced them. In fact, this approach proved to be productive, yielding results that were often unexpected and counter-intuitive.

That approach is not without its difficulties. Comparing the strategic importance of aircraft that performed different missions in different theaters at different times poses obvious problems. The fact that aviation technology changed enormously during the period of our concern further complicates matters. In addition, we must consider counterfactuals if the inquiry is to make sense. My rankings are thus indicative rather than definitive. Still, I am satisfied that the rank-ordering reflects strategic reality. I could easily justify moving many of the aircraft on the list up or down several places, but I am confident that the ranking is an accurate—albeit inexact—measure of relative strategic importance. To establish the ranking, I approached aircraft that played a major operational role in World War II with two questions: How did the aircraft in question strategically affect the conduct and outcome of the war? How would the conduct and outcome of the war have changed if the aircraft in question had not been developed and produced?

Neither question can be answered in any definitive sense. This is particularly true of the second question, which requires us to consider the responses of historical actors to events that did not, in fact, transpire. But while the answers may not be definitive, asking the questions enhances our understanding of both the design process and the nature of World War II. You, the reader, must judge the value of the project.

To keep this work to a reasonable length, I truncated my analysis, focusing on the American experience. That proved to have value in its own right, highlighting the relationship of the American aviation industry to the armed forces and government of the United Sates and how that relationship differed from those prevailing among the European powers and in Japan.

Before presenting the list, a few points about the ranking process are in order. First, the rankings are heavily—though not exclusively—dependent on chronology. The circumstances of each successive campaign and battle were determined by those that went before, so aircraft whose strategic importance was manifested early in the war generally rank ahead of those that appeared later. German victory in the Battle of France determined that the war would be a long one if it did not end in outright Nazi victory, so the aircraft instrumental in the defeat of British and French forces in May and June of 1940, the Messerschmitt Bf 109 and the Junkers Ju 87, go to the top of the list. That those same aircraft were instrumental in the early German victories on the Eastern Front reinforces their position; so does the fact that the Bf 109 in its later versions played a preponderant role in defending the Reich against daylight attack.

Victory in the Battle of Britain was an essential precondition for eventual Allied victory, so the fighters responsible for turning back the *Luftwaffe* in the summer and autumn of 1940, the Hurricane and Spitfire, come next, and so on.

Next, to say that the course of the war would have differed significantly had a particular aircraft not been designed or produced implies that there was no available substitute. The Focke Wulf Fw 190 does not make the list for this reason, reinforced by the fact it did not enter service in significant numbers until early 1942. As good a fighter as it was, most of the strategic benefits it bestowed on the Reich could have been obtained by increasing Bf 109 production. Conversely, in the strategically decisive struggle for air supremacy over Germany from late 1943 on, the Bf 109 *could* do one essential thing that the Fw 190 could not: survive in air-to-air combat against P-47s, P-38s, and P-51s at altitudes of 25,000–30,000 feet. I applied the logic of this example throughout in determining which aircraft to include or exclude from my short list.

In a few cases, the strategic impact of a given aircraft was so great as to justify moving it higher than the timing of its operational debut would indicate. The B-17 is the salient example of both this point and the previous one. The rationale behind my decision to rank the B-17 as I did, fifth on the list, is thus worth examining in detail as an illustration of the process.

It is difficult to imagine the effective destruction of the *Luftwaffe* fighter arm prior to D-Day without the threat that high-altitude daylight precision bombardment posed to the German war economy. The German high command could concede control of the air on the Eastern Front, albeit selectively, and did so following the failed July 1943 Kursk offensive.[12] It could concede control of the air in the Mediterranean, and did so following the Anzio invasion. It could not concede control of the daylight skies over the Reich without courting disaster. Forced to give battle over the Reich, the accumulative and synergistic effects of the Eighth Air Force bomber and fighter commands' combined combat effort reduced the *Luftwaffe* fighter arm to ruin.

A direct product of the B-17's ability to penetrate German airspace in massed formations, hit its targets with useful accuracy, and do so without prohibitive losses, forced the *Luftwaffe* to meet that fatal challenge. Ultimately, the provision of long-range fighter escort enabled strategic bombers to accomplish their designed mission—high-altitude daylight precision bombing. The only available substitute, the B-24, was a useful supplement to the B-17, but had to be employed with circumspection in a high-threat environment. With a service ceiling some 5,000 feet lower than that of the B-17,[13] the B-24 was considerably more exposed to anti-aircraft artillery, a liability multiplied by the B-24's greater vulnerability to battle damage.[14] Moreover, the B-24 was significantly more difficult to fly. The problem was particularly acute in the earlier versions and made the assembly of large formations above the undercast after individual instrument takeoffs difficult and at times impossible. As a concrete example, the B-24 equipped 2nd Bombardment Division tasked to participate in the 14 October 1943 Schweinfurt raid, managed to assemble only 21 of 58 bombers launched, too small a formation to be tactically viable, and the force diverted to a diversionary raid.[15] In short, the B-17 could have done the job alone. The B-24 could not have.

At this point, I will present my rank-ordering accompanied by a skeletal rationale for each aircraft's place within it followed by a brief discussion of the way in which the design of the aircraft in question contributed to its strategic significance. These discussions must be preceded by the caveat that in many cases we know little about the design process beyond what we can infer from physical characteristics, performance data, and the operational record. I do not pretend that the ranking is definitive and have no doubt that it will be controversial. It does, however, raise important questions concerning aircraft design and how it was turned to strategic advantage—or disadvantage—that we will address in concluding.

World War II Aircraft in Order of Strategic Importance

The Messerschmitt Bf 109

The Messerschmitt Bf 109[16] provided the battlefield air superiority essential to German victory in the Battle of France and the initial successes on the Eastern Front that inflicted horrendous losses on Soviet forces and materially lengthened the war. With Germany on the defensive, the Bf 109 was the *Luftwaffe*'s most important daylight interceptor and the only one capable of contesting the high-altitude daylight skies over *Festung Europa* with the long-range USAAF fighters that began penetrating German airspace from the end of 1943.

While aircraft were not designed to ideological specifications, the Bf 109 fit Hitler's strategic vision like a hand in a glove. The smallest airframe that could be built around the most powerful engine available, the Bf 109 owed its early success largely to the excellence of its Daimler-Benz 601 engine. While the DB 601's closest equivalent, the British Rolls-Royce Merlin, offered better performance at high altitudes, the German engine held a progressively greater advantage as combat altitudes dropped below 15,000 feet. This was a product of the DB 601's hydraulically driven, variable speed supercharger. The Merlin's supercharger had a mechanical clutch; it therefore ran full speed or not at all, and engaging it at too low an altitude would overboost the engine. By contrast, the Bf 109's supercharger gradually throttled back as altitude decreased and continued to yield the maximum boost that the engine could absorb right down to the deck. Moreover, the Bf 109E, the principal version employed in the Battle of France and the Battle of Britain, had the most effective armament of any contemporary operational fighter in the form of two wing-mounted high velocity 20 mm cannon, supplemented by two 7.92 mm machine guns mounted in the engine cowling. The Bf 109's cannon yielded major tactical advantages over machine-gun armed opponents, particularly in fighter-versus-fighter combat. Not only did each round inflict far more damage, the destructive effect did not diminish with range. Of considerable operational importance to the early German victories, ground crews easily maintained the Bf 109 in the field: an engine change could be accomplished in fifteen minutes. Finally, the simple and efficient design was well suited to mass production.

The only putative alternative to the Bf 109, the Heinkel He 112, had a heavier airframe and, in its initial versions, inferior flight characteristics. It would have been more difficult to maintain in the field. Finally, its more complex structure would have been more difficult to produce, the factor that ultimately led to its rejection. Significantly, the decision to reject the He 112 in favor of the Bf 109 was made in 1936 within the inner circles of the Nazi Party under the pressure of Hitler's strategic agenda.[17]

An important component of the Bf 109's early successes was the development in the Spanish Civil War of tactics based on the use of air-to-air voice radios that enabled element leaders to rely on wingmen to cover their tails and gave formation leaders a means of coordinating attacks. Called "finger four" because the spacing of the four fighters in the basic *schwarm* formation resembled that of the tips of the fingers of an outstretched hand, these tactics were later widely imitated, but gave the *Luftwaffe* an enormous initial advantage.

On the down side, a light and simple design gave the Bf 109 exceptional performance at the expense of a restricted radius of action and entailed compromises in handling characteristics. The Bf 109 lacked a rudder trim that could be adjusted in flight, placing significant demands on pilot strength and skill. The main landing gear, attached to the fuselage rather than the wing, permitted a lighter structure, but was inherently weak and placed the main wheels close together near the center of gravity. As a result, unless in the hands of an experienced pilot, the Bf 109 was susceptible to ground looping during takeoff or landing roll. In such an event, the landing gear was prone to collapse with the aircraft rotating horizontally around the landing gear. This was a significant cause of operational losses, particularly when operating from unprepared grass strips.

The net result was an aircraft capable of controlling the airspace over fast-moving armored columns, albeit at a considerable cost in operational wastage. The underlying technological strategy—implicit, but integral to the Nazi ethos—assumed that the ensuing victories would be quick and decisive, making high loss rates acceptable since they would only be sustained for brief periods. Except for the Battle of Britain, that logic remained operationally valid through the summer of 1942. Then, with the turn of the tide of the air war, first in the Mediterranean, then in the east and finally in the west, high losses in the absence of quick victories plagued the *Luftwaffe*. Although handicapped by short range, later versions of the Bf 109 remained tactically viable until war's end and German aircraft industry produced it in greater numbers than any other World War II aircraft, more than 33,000, with the sole exception of the Soviet Il 2 *Shturmovik*.[18]

In the final analysis, the *Luftwaffe* lost the war in the air by virtue of its inability to make good the loss of skilled pilots, particularly fighter pilots. While aircraft production outpaced losses almost to the bitter end, the *Luftwaffe*'s shortsightedness in failing to field a robust and dynamic pilot training establishment kept it from absorbing operational pilot losses. Shortages of aviation fuel caused by allied bombing likewise contributed to a reduction in training tempo. As a consequence, by 1945 the *Luftwaffe* defended German skies with a reduced number of fighter pilots with less experience and inadequate flight training. Pilot losses incurred as a direct product of the Bf 109's design flaws was a major factor as well. Both the failure to create a capable training establishment and the Bf 109's design deficiencies accurately reflected the *Wehrmacht*'s—and Hitler's—strategic mindset.

The Junkers Ju 87

The Junkers Ju 87's combination of bombing accuracy and psychological shock effect—an effect magnified by wind-driven sirens mounted on the landing gear and "screamers" on the bombs—made essential contributions to German victory in the Battle of France and to German ground offensives on the Eastern Front through the summer of 1942. Although not employed in a true close air support role, it provided the mobile heavy artillery that the Panzer divisions lacked. It proved highly effective in attacks on ships, inflicting major losses on the Royal Navy in the Battle of Crete and on convoys on the Murmansk run. The *Stuka* performed as a divisional organic mobile artillery reserve—stacked at altitude over the armored "schwerpunct" through the French lines, formations of Stukas provided what we would call second echelon interdiction in highly effective manner and permitted the armored thrusts to exploit timely breakthroughs.

As with the Bf 109, the Ju 87 *Stuka*—from *Sturzkampffleiger,* diving battle plane—was tailored for Hitler's strategic vision. Doctrinally, the *Stuka* exaggerated the blitzkrieg tempo of an armored paralysis by adding flexible and organic firepower to the German mechanized ground maneuver force. Supremely effective in placing heavy ordnance precisely on target, it was the only World War II bomber capable of attacking in a true vertical dive with all the advantages in accuracy that entailed. That ability played large in the Battle of France and German victories on the Eastern Front through the summer of 1942. But a true vertical attack and the high "g" forces sustained in recovery called for an exceptionally robust and heavy airframe and that, in turn, compromised maneuverability and speed. The *Stuka*'s exceptional accuracy and bomb carrying capability thus came at a price: it was horribly vulnerable to fighter attack, a lesson the *Luftwaffe* learned to its chagrin in the Battle of Britain, when *Luftwaffe* chief Hermann Goering elected to take it out of action after devastating losses. Although no longer viable in the west, the Ju 87 continued to play a useful role on the Eastern Front to the end of the war both as a dive-bomber and as a "tank buster" with a pair of 37 mm cannon mounted beneath the wings.

The Hawker Hurricane

Designed as part of an integrated, radar-controlled, air defense system, the Hawker Hurricane was essential to British victory in the Battle of Britain. Intended to bring firepower to bear against bomber formations, the Hurricane area interceptor acted in concert with its Spitfire sister, the better air superiority interceptor of the two.

A competent, workmanlike, design, the Hurricane was a straightforward development of the Hawker Fury biplane fighter, built with a traditional structure (in the early models only the fuselage from the cockpit forward had an aluminum skin) that lent itself to mass production and easy repair. It derived its tactical effectiveness from the excellence of its Rolls-Royce Merlin engine, about which an additional word is in order. A product of the Rolls-Royce company's systematic development of high-performance liquid-cooled V-12 engines that went back to World War I, the Merlin was a scaled-up development of the Kestrel that powered the Fury. The Merlin's design was also influenced by the Rolls-Royce "R" racing engine of 1931, the product of a government-subsidized program to compete in the Schneider Trophy seaplane races. The Merlin's debt to the "R" included the adoption of American-developed 100-octane fuel (about which more below) and a mechanical supercharger of unprecedented efficiency that give the Merlin exceptional performance at altitudes from 15,000 feet up.[19] The firepower of the Hurricane's massed battery of eight, and in later versions twelve, wing-mounted .303 caliber machine guns, contributed to its aerial successes against *Luftwaffe* medium bomber formations.

The Hurricane's strategic effect as an interceptor was a product of its design specification, one that called for a high rate of climb and heavy firepower at the expense of loiter time and radius of action.[20] Integral to the Hurricane's design and to its success as an interceptor was its use in ground-controlled radar intercepts directed by voice radio. Serendipitously, the peculiar circumstances of the Battle of Britain, fought at altitudes of 20,000 feet and above where the Merlin outperformed the DB 601 mitigated the Hurricane's tactical liabilities *vis-á-vis* the Bf 109. Moreover, the Hurricane's .303 caliber armament, inferior to the Bf 109's cannon in fighter-versus-fighter combat, proved brutally effective at close range against lightly armed and armored *Luftwaffe* bombers. Finally, as already noted, the Hurricane was eminently producible. In addition to its key role in the Battle of Britain, the Hurricane performed yeoman duty as a fighter-bomber in North Africa and when deployed aboard merchant ships as expendable catapult-launched interceptors Hurricanes helped counter the depredations on Allied shipping inflicted by Focke Wulf Fw 200 reconnaissance bombers that reached epidemic proportions in the final months of 1940. Carrier-based Sea Hurricanes played a small, but significant, role in the war at sea.

While the Hurricane's usefulness as a ground attack fighter bomber was mitigated by a short radius of action and the inherent vulnerability of liquid-cooled engines to battle damage—a single hole in the coolant system can drain the engine, leading quickly to seizure—it was the best the RAF had until the advent of the Hawker Typhoon, which shared the same liability. Fitted with intake filters to protect the engine from sand particle erosion and armed with a variety of wing- and under-wing-mounted cannon, the Hurricane was effective in the ground attack role in North Africa. The Supermarine Spitfire had significantly better performance and would in principle have been available as a substitute, but was significantly harder to produce and repair. In consequence, it is doubtful that enough Spitfires could have been built and kept in commission to defeat the *Luftwaffe* in the Hurricane's absence. The figures bear this out: during the Battle of Britain the number of Hurricane squadrons increased from twenty-five to thirty-three while the number of Spitfire squadrons remained constant at nineteen.[21] Given inspired leadership—which Fighter Command had—Britain could probably have prevailed in the Battle of Britain without the Spitfire. British victory is difficult to imagine without the Hurricane.

The Supermarine Spitfire

Like the Hurricane, the Supermarine Spitfire was procured as part of an integrated, radar-controlled air defense system. Capable of meeting the Bf 109E on an even footing at 15,000 feet and with an increasing advantage as altitudes rose, it made a vital contribution to British victory in the Battle of Britain. Fighter Command employed Spitfires to defeat the *Luftwaffe* fighter escorts while Hurricanes attacked and shot down the *Luftwaffe* bombers. From the end of the Battle of Britain until late 1943, it was the only Allied day fighter available in numbers in the European theater that could match the performance of first-line German fighters and remained in front-line service until the end of the war. During 1941–42, Spitfires played a major role in wresting air superiority over North Africa and the Mediterranean from the *Luftwaffe* and the Italian *Regia Aeronautica,* a matter of no small strategic importance. In addition, specially modified Spitfires were the most important Allied strategic photo-reconnaissance aircraft at the outbreak of hostilities in 1939—and the *only* ones capable of deep penetrations of Axis territory—and so remained until the debut of reconnaissance versions of the Mosquito in the autumn of 1941.[22] The Spitfire remained effective and important in that role until war's end. As with the Sea Hurricane, Seafires, as the carrier-based Spitfire was called, played a small but significant role in the war at sea.

Designed to a specification that called for a high rate of climb and heavy firepower at the expense of range and loiter time, the Spitfire, and, like the Hurricane, owed its tactical success to its Rolls-Royce Merlin engine fueled with 100-octane aviation gas. The Spitfire too was a lineal descendant of the Rolls-Royce "R" powered Supermarine S 6 racer that won the Schneider Trophy in 1931. To an even greater extent than the Hurricane, the peculiar circumstances of the Battle of Britain negated the Spitfire's tactical liabilities. More aerodynamically refined than the Hurricane—its elliptical wing planform increased aerodynamic efficiency by some 2 to 3 percent.[23] Inferior to the Bf 109E at low altitudes, the Spitfire Mark II enjoyed appreciable advantages in maximum speed and turn radius at the altitudes at which the combatant pilots fought the critical engagements of the Battle of Britain. As the war progressed, the Spitfire was given remarkable longevity as a first-line air-to-air fighter by progressive improvement of its Merlin engine, but the Spitfire was more difficult to produce and repair.

Comparison with the Daimler-Benz powered Bf 109 is instructive in this regard. While early versions of the DB 601 were superior to contemporary Merlins in power to weight ratio and in performance at medium and low altitudes, the DB 601 proved unable to accept increases in compression ratio that the more solidly built Merlin absorbed with ease. In consequence, the G model of the Bf 109, fielded from the summer of 1942 and the most important variant in numbers produced, required an entirely new engine, the DB 605.[24] Moreover, while engineers increased the Spitfire's armament from late 1940 to include two wing-mounted 20 mm cannon, and later four, wing-mounted cannon had to be abandoned on Bf 109s beginning with the F model in 1941 to avoid fatally compromising performance in air-to-air combat. The final production versions of the BF 109 had only a single cannon firing through the propeller hub supplemented by two 12.7 mm machineguns in the engine cowling. The British parallel to the DB 605 was the Rolls-Royce Griffon, a development of the "R" racing engine that was begun in 1933, then set aside until 1939.[25] Like the DB 605, the Griffon developed more power than its predecessor in the same space, 2,035 to 1,700 horsepower in late 1943 versions. Unlike the DB 605, the Griffon was strategic insurance rather than a necessity. While Griffon-powered Spitfires and Seafires were tactically superior to Merlin-powered versions, the latter remained tactically viable. Only in photo-reconnaissance Spitfires did the Griffon's added power and efficiency yield strategically important dividends.

The Griffon-powered Spitfire PR XIX (PR for photo-reconnaissance), which entered service in the spring of 1944, provides a final commentary on the Spitfire's importance. The Griffon's superior high-altitude performance and a pressurized cockpit combined with the Spitfire's refined aerodynamics to give the PR XIX a service ceiling of no less than 48,000 feet—the highest of any operational piston-engined aircraft—rendering it effectively immune from interception. At that point photo-reconnaissance versions of the P-38 were horribly vulnerable to interception by later versions of the Bf 109 and the PR XIX, though produced only in small numbers, satisfied a vital strategic requirement at a critical time.

The Boeing B-17

The Boeing B-17 Flying Fortress was the anvil against which the USAAF fighter force hammered the *Luftwaffe* fighter arm to destruction in the skies over Germany. That was of immense strategic importance above and beyond the destruction that B-17s visited on military and industrial targets, threatening a level of damage to key industries that the Third Reich's leaders could not tolerate. The B-17 was a singular design for which there would have been no viable substitute until the B-29 became available in quantity . . . if Boeing could have designed the B-29 without the experience gained from the B-17. Even before long-range fighters were available to escort deep penetrations, massed formations of B-17s took a significant toll of the *Luftwaffe* fighter arm, both physically and psychologically, helping to make subsequent the German air arm's recovery impossible.

The B-17 was an uncompromising 1934 design intended to produce the fastest, highest-flying heavy bombardment aircraft extant. Boeing's design team adopted those objectives in response to stated Army Air Corps requirements, but pushed them to the limit as a conscious high risk, high gain strategy to deliver blows against an industrial enemy. For a variety of reasons involving internal Army politics and blind luck—the loss of the first prototype to a pilot error accident—that strategy nearly failed and most of the initial Army bomber contract went to the mediocre twin-engine B-18, a derivative of the DC-3 civilian transport. That having been said, Boeing's boldness reaped huge strategic dividends in range, bomb load and ability to absorb battle damage.

The excellent Wright R-1820 nine-cylinder engine, reengineered at Air Corps insistence to burn 100-octane gasoline, was an essential cornerstone of the B-17's success.[26] Another was the development of the turbo supercharger by General Electric on an Air Corps contract, the only discrete Army research and development program to receive funding through the Great Depression. The importance of the turbo supercharger lies in the fact that the War and Navy departments stopped subsidizing the development of military aero engines during the Great Depression. American military aircraft would henceforth be powered by engines designed for civilian use, and while high-altitude performance had obvious military importance it had little civilian value. The European solution, gear-driven superchargers designed as an integral part of the engine, was an obvious non-starter for economic reasons.[27] The military market was simply too small. American superchargers therefore would be add-on accessories and the only evident way to power such a supercharger was a turbine driven by engine exhaust gasses. The extremely high temperatures and rotational speeds to which the turbines were subjected posed obvious problems. A further complicating factor was the lack of full-sized high-altitude wind tunnels: turbo superchargers could only be tested in actual flight with the obvious risks that entailed. Beginning work in 1919, General Electric eventually surmounted these problems and by the mid-1930s was fielding increasingly reliable turbo superchargers.[28] The B-17 was slated for them from the beginning.

The result was a bomber capable of delivering a two-ton bomb load over a thousand miles from its base—the figures are approximate, based on data from missions flown over Germany in 1943–44—penetrating enemy air defenses *in formation* at altitudes of 25,000 to 29,000 feet.[29] The emphasis in the preceding sentence is warranted since a formation's speed and ceiling are dictated by its most poorly performing aircraft. Such performance, unprecedented in the mid- to late 1930s, speaks volumes both for the soundness of the B-17's design and for the excellence of Wright, General Electric, and Boeing production line quality control. The excellence of the B-17's design is highlighted by comparing it to that of the Consolidated B-24 *Liberator,* the closest thing to an available substitute. A newer design by five years and similarly powered,[30] the B-24 was nonetheless inferior to the B-17 in every critical performance parameter that counted in the European theater of operations save maximum range.

A final factor contributing to the B-17's success was the decision by the Air Corps during the 1920s to adopt the .50 caliber machine gun as its standard aircraft weapon. Designed toward the end of World War I as a heavy infantry machine gun, the Browning .50 caliber was an uncompromising design with exceptional ballistic performance.[31] Not only was its projectile nearly four times as massive as that of .30 caliber weapons, its superior ballistic coefficient and streamlined shape gave it the best velocity over distance characteristics of any commonly used aerial machine gun of World War II.[32] As a result, the Fortress's effective defensive fire ranged well beyond the practical hitting distance of any Axis air-to-air gun. While unescorted B-17 formations proved unable to sustain deep penetrations of German airspace without incurring prohibitively heavy losses, they inflicted serious losses on the German fighter arm in the process. To be sure, the Air Corps initially underestimated the need for defensive armament and Boeing engineers resisted the addition of turrets that spoiled the aircraft's aerodynamic shape. Ultimately, however, tactical logic and superior engineering prevailed and from early 1942 on B-17s were well provided with heavy defensive armament, much of it mounted in power-operated turrets.

Facilitated by intercom and radio connections, the Fortress' aircrew arrangement throughout the cockpit, crew compartment and fuselage ensured that the dispersed crewmen retained their group cohesion in air to air combat. Compartmentalized responsibilities and specialized training demanded aircrew discipline in coordinating defensive fire and fighting battle damage to the airframe, engines, and subsystems. This dispersed crew arrangement provided for more defensive armament that could protect the bomber from all flight attitudes of fighter attack, especially with the B-17G modified "chin" turret model giving frontal attack defense. The size of the airframe and engineering capacity enabled the Fortress to grow in defensive firepower from ten .50 caliber machine guns aboard the "E" model to the "G" model with thirteen machine guns providing all around defense. The porcupine firepower gave large formations of B-17s overlapping fields of fire that enabled adjacent elements and squadrons to cover one another. The result was the fifty-four aircraft combat box formation that dealt severe blows against the industrial strength Nazi Germany.

The final analysis, bombers exist to drop bombs, and a late war USAAF study showed that the B-17 was the most accurate Army bomber (the second most accurate being the B-29), enjoying a small, but significant, advantage over the B-24 despite the fact that B-24s bombed from lower altitudes.[33] In *ex post facto* validation of the Air Force's preference for heavy bombers, the study showed that four-engine bombers were significantly more accurate than twin-engine bombers across the board.[34] That the B-17, a 1934 design, was still in front line service in 1945 speaks volumes for the quality of its design.

The AVRO Lancaster

The AVRO Lancaster was the backbone of RAF Bomber Command's night area bombardment campaign, and while that campaign failed to defeat the Third Reich in isolation, it wrought immense destruction, forced strategically important diversions of resources, and—a critical point often forgotten—was Britain's only means of taking the war directly to Germany until D-Day. Without the Lancaster, it is unlikely that the night area bombardment campaign could have been sustained during 1943–44 without unbearable losses. The only available substitute, the Handley-Page Halifax, was a far less capable aircraft with a much lower service ceiling and a significantly higher loss rate: Lancasters dropped 107 tons of bombs for every one lost in combat,

Halifaxes only 48. Moreover, the Halifax was more difficult to produce and maintain, consuming 11,000 man-hours of labor per ton of bombs dropped to 4,000 for the Lancaster.[35] Lancasters also made major contributions to the preparations for D-Day and to the destruction of the German oil production in 1944 and German rail transportation net during the winter of 1944–45. In addition to its positive contributions to Allied victory, the Lancaster program absorbed an immense quantity of vital resources, a matter of considerable strategic significance.

The Lancaster was a derivative of the AVRO Manchester, a heavy night bomber designed to a 1936 contract and powered by two Rolls-Royce Vulture engines.[36] By mating two V-12 Merlin equivalents belly-to-belly around a common crankshaft,[37] the Vulture doubled the engine power output while halving the number of engine nacelles, thus reducing aerodynamic drag. As with virtually all liquid-cooled in-line engines having more than twelve cylinders, the Vulture suffered extensive development problems[38] and, though these were eventually technically solved, the Manchester still suffered from being badly overweight.[39] RAF Bomber Command consequently withdrew the Manchester from operations after a brief career, but the airframe showed promise and a substantial investment had been made in production facilities. In an inspired decision to salvage the investment, four Merlin engines, mounted on a larger wing; a redesigned empennage and a name change to Lancaster produced a strategic bomber capable of carrying the largest possible bomb load at medium altitudes better than any other World War II bomber. Beyond gross bomb carriage capacity, the Lancaster was designed and modified to carry an unprecedented variety of bombs, ranging from 4 lb incendiaries through conventional high explosive 500 lb and 1,000 lb bombs and the 4,000 lb light case "blockbuster" to the 12,000 lb Tallboy and 22,000 lb Grand Slam. The small incendiaries were particularly effective in attacks on oil refineries when used in combination with high explosive bombs and played a significant role in the strategically decisive 1944 campaign against German oil. The Tallboy and Grand Slam, though not available until 1944, proved devastatingly effective in the 1944–45 campaign against German transportation. The Lancaster's effectiveness as a bombing platform was multiplied from 1942 by the development of effective blind-bombing aids.

Against the Lancaster's unparalleled ordnance carriage capabilities, the design tradeoff was a modest service ceiling of around 24,000 feet that made daylight operations infeasible except in the most permissive of environments and calls into question the prewar RAF's appreciation of the lethality of German anti-aircraft artillery. The RAF's reliance on .303 caliber machine guns for defensive armament further constrained the Lancaster's effectiveness. These were badly outranged by the high velocity 20 mm cannon carried by *Luftwaffe* night fighters, all but reducing British gunners to lookouts. To compound matters, the British did not field an effective belly turret and Bomber Command decided to eliminate downward firing armament altogether on the mistaken assumption that *Luftwaffe* night fighters would not attack from that quarter. In fact, the *Luftwaffe* installed upward-firing cannon in its night fighters, mounted in the fuselage and firing forward at a 10°–20° angle from the vertical so that the pilot could aim by means of a sight mounted in the top of the canopy. These went undiscovered for an extended period and inflicted heavy losses on Bomber Command. In combination with American long-range daylight bombers the Lancaster forced dispersal of the German armaments factories, forced Germany to deploy a million man air defense force to protect the homeland, diverted German industrial production from offensive weapons and contributed to the effects of attrition on the Eastern front.

The Mitsubishi A6M Zero

The Mitsubishi A6M Zero was the linchpin of early Japanese strategic success. Without the Zero's range and effectiveness in air-to-air combat, the Pearl Harbor attack and the conquest of the Philippines and Netherlands East Indies would have been problematic at best. The Zero was an improbably good design, and one for which there was no available substitute. On the negative side of the strategic ledger, the Zero's remarkable performance was gained at the expense of vulnerability to battle damage. Its tactical effectiveness was thus heavily dependent upon pilot skill, magnifying the strategic impact of the loss of the Japanese Navy's cadre of experienced aviators in the Solomons campaign.

A combat aircraft designed to a tight and seemingly impossible specification calling for unprecedented range and maneuverability in a carrier fighter, the Mitsubishi A6M Zero is the rare example of a first-rate combat aircraft powered by a mediocre engine. Indeed, Japanese engineers consciously compensated for the fact that Japanese aero engines were, quoting the Zero's designer Horikoshi Hiro, "20 to 30 percent less powerful than those of the more advanced countries."[40] That notwithstanding, the Zero was the first carrier-based fighter capable of besting its land-based equivalents. This is remarkable in light of the fact that the design of carrier-based aircraft is inherently more difficult than that of the land-based equivalents. Not only do arrested carrier landings call for a considerably stronger, and hence heavier, structure; final approach speeds must be low by land-based standards and handling characteristics must be exceptionally good if high operational losses are to be avoided. The Zero's range, an essential precondition to early Japanese victories in the Pacific, was the compromise of an extremely light, yet strong, structure and the provision of a jettisonable centerline external fuel tank. The Zero's remarkable maneuverability in air-to-air combat combined a low wing loading and excellent power-to-weight ratio[41] with a potent armament of two wing-mounted 20 mm cannon plus two 7.7 mm machine guns in the engine cowling, mainly to help the pilot aim the cannon. In order to obtain the remarkable wing loading and power-to-weight ratio that made the Zero formidable, designer Horikoshi dispensed with protective armor and self-sealing fuel tanks and Zero pilots wore no parachutes. This was not, as is commonly imputed, because the Japanese Navy placed a low value on the lives of its pilots or because of a "kamikaze mentality," but due to a rational assessment of pilot survival factors. Unlike its main allied opponents, the Zero, with flotation bags in the wings, had excellent ditching characteristics.[42]

The Zero's critical dependence upon pilot skill was its Achilles heel. Once the Japanese Navy had expended its cadre of skilled aviators in the Solomons campaign, the Zero's prime liability, extreme vulnerability to battle damage, made it a death trap.

The F4F Grumman Wildcat

The only battle-worthy American fighter in operational service in 1941, the F4F Grumman Wildcat assumed strategic importance by virtue of its ability to take the Zero's measure. This had two principal strategic effects, one intangible, the other attritional and both of great importance. Though the evidence is circumstantial, it is clear that confidence in the Wildcat and the men who flew it emboldened our naval commanders to challenge the Japanese aggressively in the early days of the war. This led to victory at the Coral Sea and Midway. Second, the Wildcat played a dominant role in the destruction of the flower of the Japanese naval air arm in the Solomons campaign, particularly in the critical early stages. That the Wildcat's tactical effectiveness was largely due to remarkable prewar tactical innovation within the U.S. Navy fighter community in no way lessens its strategic importance. The Brewster F2A Buffalo, the only putative substitute until the operational debut of the Grumman F6F Hellcat in August 1943,[43] was a deathtrap. A superior design, the Hellcat added significantly to Japanese losses and lessened American casualties, but entered service only after the Japanese naval air arm had been effectively destroyed.

Like the Hurricane, the Wildcat was the lineal development of a biplane precursor, the F3F, and was a conservative design structurally. The Wildcat owed its robust performance to its Pratt and Whitney R-1830 fourteen-cylinder, twin-row radial engine, fitted with a two-stage, two-speed mechanical supercharger in the initial operational versions. So powered, the Wildcat was inferior to the Zero in turn radius, rate of climb and climb angle, deficiencies that should have placed it at a severe tactical disadvantage. It could match the Zero in service ceiling and—almost—in maximum speed in level flight. It could easily outstrip the Zero in a dive. Given the Japanese fighter's vulnerability to battle damage, the Wildcat's four .50 caliber machine guns were a match for the Zero's comparatively low-velocity 20 mm cannon. With shorter wings, the Wildcat also had a higher initial roll rate, essential for breaking contact with an enemy on your tail. Finally, with self-sealing tanks and armor protection for the pilot, it was far more resistant to battle damage and lessened pilot combat attrition meaning more American pilots survived lost engagements to become battle hardened and experienced for their next aerial combat.

The Wildcat's performance might have gone for naught had the U.S. Navy's fighter community not developed remarkably innovative tactics during the late 1930s. First, the U.S. Navy's air service was effectively alone among the world's air forces in systematically training its fighter pilots in wide off-angle deflection shooting, meaning that they were trained to lead their targets by as much as 60°.[44] Second, eschewing the then-*de rigueur* three-ship "Vee" and echelon formations prevalent in every air force but the *Luftwaffe,* the Navy embraced a system of mutually supporting two-ship, two-element formation tactics developed by Commander James Thatch in which each pilot in the four ship formation continually checked one another's blind spots astern, the so-called "beam defense position" or "Thatch weave."[45] The Wildcat's design serendipitously enhanced its tactical effectiveness in that the pilot sat high in the cockpit above the wing and engine, primarily for better visibility in carrier landings, and downward visibility over the nose, already good by design, was enhanced by the R-1830's small diameter. The result was an important tactical advantage: when "pulling lead" in attacking a turning enemy from astern, Wildcat pilots could maintain visual contact at closer ranges and thus press home their attacks more aggressively than could their Japanese opposites, seated low behind the Zero's larger engine.

The Douglas SBD Dauntless

The Douglas SBD Dauntless' strategic importance derives first and foremost from the destruction of the heart of the Japanese fast carrier force in the Battle of Midway. Victory at Midway precluded a massive redeployment of American resources to the Pacific that would have undercut the Allied Europe First strategy and lengthened the war by six months to a year. The Dauntless also played a pivotal role in the Guadalcanal campaign, blunting the power of the Japanese Navy when it still enjoyed a measure of operational freedom, wreaking havoc on Japanese warships and shipping. The Dauntless remained the Navy's principal dive-bomber until well into 1944 and accounted for a greater tonnage of Japanese warships sunk than any other American aircraft.

Designed for one thing and one thing only, the destruction of enemy warships, the Douglas Dauntless was a less efficient dive bombing platform than the Ju 87, but a far superior aircraft in every other regard. Edward Heinemann, although not formally trained as an aeronautical engineer, conceived of a light, strong airframe, first incorporating it into the wing for the Douglas DC 3, and then into the Northrop A-17 attack aircraft. He extrapolated this sturdy and light design into the SBD. Underpowered, the SBD had a low rate of climb with a full bomb load and was not particularly fast, but had sterling flight characteristics in all other respects and was an excellent instrument platform. Of considerable importance, its carrier landing characteristics were excellent, a fact that reduced operational wastage.

The Republic P-47 Thunderbolt

The Republic P-47 Thunderbolt was the first American fighter available in significant numbers in the European Theater that was capable of reaching German airspace and out-performing first line *Luftwaffe* fighters at high altitudes. That made it possible for USAAF heavy bomber formations to attack targets inside Germany without prohibitive losses and forced the *Luftwaffe* fighter arm to accept battle on unfavorable terms, leading ultimately to its defeat. The Lockheed P-38, with a significantly greater radius of action than the P-47, entered operational service earlier and could have done the job in principle, but was never wholeheartedly embraced by the USAAF and was not available in quantity at the critical time. The North American P-51 could have done the job as well and ultimately did, but became available in numbers only after the

P-47 had turned the tide. Produced in larger numbers than any other American fighter, 15,579, the P-47 was a highly effective as a fighter bomber in the European Theater and played a major role in interdicting German lines of communication and in supporting friendly ground forces.

The P-47 was the end product of an evolutionary series of fighter designs by Russian émigré designer Alexander Kartveli that combined a powerful radial engine, an all-metal structure, and an elliptical wing in the smallest airframe feasible. Starting with the P-35 of 1935, Kartveli's fighters became progressively larger and more aerodynamically refined, acquiring a turbo-supercharger in 1939 with the R-1830-powered P-43. Within the parameters of his basic design, Kartveli perceived that the full benefits of turbo supercharging could only be realized with an engine in the 2,000-horsepower range. In 1940, he turned to the eighteen-cylinder, twin-row radial Pratt and Whitney R-2800 just entering production to harness the raw power of a turbo supercharger. The Pratt and Whitney was engineered, like all high-performance U.S. aero engines to exploit the properties of high-octane gasoline. The integration resulted in the P-47, the largest single-engine, piston-powered fighter ever built. Remarkably clean and sophisticated, the P-47 was one of the few successful mid-wing fighters of the war and, like the Spitfire, benefited from the greater efficiency of an elliptical wing.[46]

In fighter installations, the turbo supercharger offered advantages similar to those of a variable clutch mechanical supercharger that could be progressively disengaged to obtain maximum power at lower altitudes without over boosting the engine. Achieving the benefits of the turbo's inherent characteristics required superior thermodynamic-mechanical and aerodynamic engineering and the P-47's supercharger installation, although necessarily complex, was remarkably efficient and reliable. Taking full advantage of the R-2800's power, Kartevli gave the P-47 an armament of no less than eight wing-mounted .50 caliber machine guns.

The P-47 could carry a significant bomb load and that, combined with its heavy firepower and the remarkable ability of the R-2800 to absorb battle damage, made it one of the most effective fighter-bombers of the war. Of greater strategic importance, the P-47 could also carry over 200 gallons of fuel in jettisonable external tanks. In late 1943 when the USAAF came belatedly to an appreciation of the value of long-range fighter penetrations of *Festung Europa* in support of heavy bomber operations, the P-47 was the only U.S. fighter available in significant numbers that possessed the requisite capabilities. At that point, P-47s based in southeast England could penetrate only as far as an arc running through Lübeck and Frankfurt. P-38s could penetrate as far as Leipzig from the beginning and by February 1944, were equipped with larger external tanks, enabling them to reach Berlin, but USAAF production and deployment decisions limited their availability to small numbers. P-51s, capable of reaching well beyond Berlin to as far as Prague and, eventually, Vienna, only became available in significant numbers from March 1944.[47] In the meantime, the P-47 filled the gap. Had Air Force leaders appreciated the importance of long-range escort fighters sooner than they did, the P-47 could have been readily modified to match the radius of action of the P-38 and P-51. The longer-ranged Thunderbolt, in the form of the P-47N, saw action in the Pacific in the final days of the war.

The Yakovlev Yak 1–9

The first Soviet fighters capable of meeting the Bf 109 and Fw 190 on even terms, Yak fighters were produced in large numbers and played a major strategic role in denying the *Luftwaffe* the unimpeded exploitation of the air it had enjoyed to great effect until Stalingrad. The Soviet Union possessed only ninety-four Yak-1s on the eve of Barbarrossa, but by the end of the war, the rejuvenated Soviet aircraft industry had manufactured over 16,700 Yakovlev fighters, some 58 percent of all Soviet single-seat fighters, distributed among many *Protivovozdushnaya Oborona* air defense units. The Lavochkin La 5-7 series could have done the job, but entered service later, offered no tactical advantages over contemporary Yak fighters, and was not, in the final analysis, essential.

The Yak-1 was an unremarkable, straightforward, and competent design, in many ways reminiscent of the Hurricane, which it superficially resembled. The circumstances of aerial combat on the Eastern Front dictated that it would fight only at low to medium altitudes, for which its liquid-cooled Klimov V-12 engine, developed from a French Hispano-Suiza original, was more than adequate. Indeed, the Yak-3, with a reduced wingspan for greater low-altitude maneuverability, was one of the best low-altitude dog-fighters of the war, a fact attested to by *Luftwaffe* flight evaluations of captured examples. The Yaks' armament, typically a 20 mm cannon firing through the propeller hub and two synchronized 12.7 mm machine guns in the engine cowling, was somewhat lighter than that of its principal *Luftwaffe* opponents, but was adequate. Better than they needed to be, Yak fighters contributed to the aerial battle by providing an environment of air superiority that permitted the *Shturmovik* ground attack units to support the highly Il-2 mobile Soviet armored forces in their counter-offensives from Stalingrad to Berlin.

The de Havilland Mosquito

The de Havilland Mosquito's strategic importance derives in the first instance from its effectiveness as a photo reconnaissance aircraft. Effectively immune to interception by virtue of its speed and service ceiling and with a significantly greater radius of action than any competing design, the Mosquito provided Allied intelligence staffs and operational planners with information of immense value, almost all of which could have been obtained in no other way. In addition, the Mosquito made important contributions to the Combined Bomber Offensive as a bomber, particularly marking targets in the pathfinder role.[48] The Mosquito also enjoyed significant success as a low-altitude precision daylight bomber, as a night fighter, as a daylight intruder fighter, and in the maritime strike role.

The Mosquito's genesis lay in a 1935 RAF specification stimulated by reports that the Germans were building an extremely fast twin-engined bomber. It called for a bomber powered by two Rolls-Royce Merlins with a defensive armament of three .303

calibre machine guns in streamlined mounts. Geoffrey de Havilland's interest in the project and his firm's experience in building high-performance multi-engined aircraft with wood structures resulted in the Mosquito. The decision to delete all defensive armament—de Havilland's preference from the start—was made by Air Vice Marshall Wilfred Freeman, the Air Ministry official in charge of production and development, in August 1939.[49] That decision was central to the Mosquito's success.

The Mosquito was a rarity: a genuinely successful multi-role combat aircraft. Although the Mosquito excelled in its intended role (it had the lowest loss rate over Germany of any British bomber) it played a significant part in the multiple roles mentioned above. Rendered safe from interception by speed, photo-reconnaissance Mosquitoes had sufficient range to cover most of Germany from bases in the United Kingdom, and after the capture of the Foggia airfield complex in Italy in September 1943 provided coverage of the entire Third Reich. Progressive development of the Mosquito and its Merlin engines kept a step ahead of German defenses, particularly for photo-reconnaissance. The Mosquito PR XIV, which entered service in late 1943, had a fully pressurized crew compartment and a service ceiling above 35,000 feet that rendered it effectively immune to interception. Only 432 PR XIVs were produced, but they rendered strategically vital intelligence.[50] Despite the need for exotic glues and highly skilled workers, De Havilland manufactured nearly 8,000 Mosquitoes. The Mosquito's drawbacks included short airframe life in tropical conditions and the difficulty of exiting a damaged aircraft in flight.

The Consolidated PBY

The Consolidated PBY was ubiquitous as a patrol aircraft for the U.S. and Royal Navies, entering service with the latter in early 1941, well before America's entry into the war. PBY crews located the *Bismarck,* gave the Royal Navy warning of the April 1942 Japanese incursion into the Indian Ocean, located the Japanese carrier force before Midway, were omnipresent in tracking Japanese task forces and convoys in the Solomons campaign, and played a major role in the Battle of the Atlantic. Allied effectiveness in dealing with Axis naval surface forces owed much to the U.S. and Royal Navies' emphasis on patrol operations in which the PBY excelled and played a disproportionately important role.

A twin-engine flying boat of conservative design, the Consolidated PBY (Catalina in British service) entered service in 1936 and possessed unremarkable performance except in range, endurance, and handling qualities. A competent design, it was the right aircraft for the job at the right time and was procured in adequate numbers by the U.S. Navy and for British and Canadian forces. That the PBY's strategic significance was due as much to the U.S. and Royal Navies's emphasis on reconnaissance in support of the battle fleet as to the excellence of its design takes nothing away from the PBY's luster. It was slow, with a cruise speed of only 179 mph, but had a radius of action of nearly 2,000 miles and an endurance of no less than 17.6 hours.[51] Less effective as an anti-submarine patrol aircraft than the B-24 by virtue of the latter's greater speed and the ease with which it could be modified to carry electronic equipment and offensive ordnance, it was still useful in that role. Later versions were amphibians, fitted with retractable landing gear. In addition to reconnaissance, it was used for air-sea rescue by U.S. Army Air Forces in the later stages of the Pacific war.

The Douglas C-47

The Douglas C-47, "Skytrain" (Dakota in British service), was far and away the most important, and best, tactical transport and paratroop deployment aircraft of the war. Produced in large numbers, it provided the bulk of the airlift that dropped two American and one British airborne divisions behind the D-Day invasion beaches and would make the list on that basis alone. In fact, the C-47 did a great deal more, hauling key personnel, spare parts, supplies and fuel. Provided in significant numbers to the British and Soviets, it was used most innovatively and in the largest numbers by U.S. forces, but served as a potent Allied logistical force multiplier in all theaters.[52]

The military version of the 1937 Douglas DC 3, the first commercially successful airliner, the C-47 was beyond doubt the most successful tactical transport of World War II.[53] A scaled up extrapolation of the DC 2 of 1934 (itself developed from the prototype DC 1 of 1933), the DC 3 varied from its predecessors in the provision of a cabin sufficiently spacious to permit passengers to stand up and walk around. The DC (for Douglas Commercial) series of transports extracted the full benefit of stressed skin aluminum construction, the wings being particularly efficient. Designed in response to a Trans World Airlines specification that stipulated that a safe takeoff at design weight following the loss of an engine could be completed from the highest airfield served by TWA following loss of an engine, the DC 3 had adequate reserves of power and was inherently safe. Modifications for military use were minimal, the most important being the provision of easily removable (and spartan) passenger accommodations, provisions for securing heavy cargo inside the cabin, and a spacious loading door. Powered by two 1,200-horsepower R-1830 engines, the C-47 was fast for a transport with a cruise speed of 185 mph; it had a useful load of as much as 14,000 pounds and a radius of action of over 700 miles.[54] It had excellent flight characteristics and was easily maintained in the field. The rear cargo door could be opened in flight, making it far and away the best mass produced paratroop deployment aircraft of the war. Total U.S. production reached 10,926, of which 10,123 were specifically manufactured as military transports; to this we can add 6,157 built under license in the Soviet Union.[55] As a tribute to the C-47's efficiency and durability, it formed the backbone of the Soviet internal air transport network into the 1960s.

The North American P-51

Entering service after the P-38 and P-47, the North American P-51 had greater range and, partly in consequence, enjoyed a better kill ratio than either of the other two fighters. It was also the more agile of the three, a fact that its pilots exploited to great effect. From March of 1944 the P-51 broke the back of the *Luftwaffe* fighter arm, probably shortening the war and surely reducing the cost to the United States.[56] The P-38 *could* have done the job—the P-47's range was inadequate—and given time would have, albeit at greater cost in blood and treasure. The P-51 *did* it. Finally, the photoreconnaissance version of the P-51, the F-6, was the most successful low-altitude photographic imagery collector of the war.

The most successful long-range piston-engine air-to-air fighter of the war, the North American P-51 was one of a handful of strategically significant aircraft to be designed after the commencement of hostilities in 1939. Ironically, the P-51's designer, Edgar Schmued, was Austrian by birth and did his engineering apprenticeship in Germany.[57] Designed in response to a request by the British purchasing commission in 1940, the North American P-51 (Apache in British service) benefited from the most recent NACA drag reduction and airfoil data, a fact that put it in a class by itself. The efficiency of the P-51's laminar flow wing and engine installation resulted not only in an excellent turn of speed, but in dramatically greater range than existing single-engine fighters. The exceptionally compact engine installation and associated engine cooling resulted in remarkably low drag. The P-51's principal liability in its early production versions stemmed from the altitude limitations of its Allison engine. Virtually identical to the Rolls-Royce Merlin in configuration, size, and development potential, the Allison was anemic by comparison due to the low rated altitude of its mechanical supercharger, the product of an Air Corps decision to rely on turbo supercharging for high-altitude performance. In the event, shortages of the high temperature alloys needed for turbo supercharger turbine blades initially limited production to little more than that required for heavy bombers and the P-38 was the only fighter powered by turbo-supercharged Allisons.

The RAF began taking delivery of Apaches in November 1941 and used them for long-range low-altitude operations from July of 1942 with considerable success.[58] Impressed by the aircraft's performance, the British experimentally re-engined an Apache with a Rolls-Royce Merlin in the spring of 1942. The results were spectacular, yielding outstanding high-altitude speed and range, but might have lead to nothing had not USAAF Major Thomas Hitchcock, assigned to the American Embassy as an attaché, been invited to fly the aircraft.[59] An instant convert, Hitchcock was both persuasive and well connected. Re-engined with the Merlin, providentially produced in the United States under license by Packard, the Mustang became the best long-range, high-altitude fighter of the war to be manufactured in large numbers, 14,819 by the end of the war. The P-51 was not without vices. The vulnerability of its liquid coolant system limited its usefulness in the ground attack role and careful management of the fuel system was necessary to avoid exceeding the rearward center of gravity limit with a full fuel load. On balance, however, it was a remarkably well designed aircraft that exceeded all expectations.

The Lockheed P-38 Lightning

The Lockheed P-38 Lightning rendered U.S. air superiority in the Pacific unassailable from the autumn of 1942 and made significant contributions to the defeat of the *Luftwaffe* in the skies over *Festung Europa* in the spring of 1944. The first USAAF—or any other—fighter capable of high-altitude escort operations deep within enemy territory, the P-38 was not used effectively in that role until the P-47 was available in larger numbers and was then superseded by the faster, more maneuverable and longer-ranging P-51. For the USAAF, the P-38 was the greatest missed strategic opportunity of the war. Because it entered operational service nine months later than it should have, because it was produced in smaller numbers than any other battle-worthy USAAF fighter, because few pilots in the European Theater learned to exploit its peculiar strengths, and because it was belatedly employed in the long-range escort role, the P-38's strategic impact was substantially less than it might have been. The photo reconnaissance version of the P-38, the F-5, though inferior to the Mosquito in range, speed and service ceiling, made significant contributions to Allied victory.

Lockheed's Clarence "Kelly" Johnson designed the P-38 in response to a January 1937 Army Air Corps' specification for a long-range interceptor so demanding as to deter other would-be contractors.[60] Perceiving that the required speed, climb and service ceiling could not be met by orthodox means, Johnson turned to a radical twin-engine design powered by a pair of turbo-supercharged twelve-cylinder, liquid-cooled Allison V-1710 engines. A central fuselage-pod housed the cockpit situated between the engines mounted in mid-wing booms that supported the tail surfaces. This configuration offered the significant ancillary benefit of grouping the armament of four .50 caliber machine guns and a 20 mm cannon closely together in the nose. Unlike wing-mounted guns that were "harmonized" so that their fire converged at a predetermined distance and dispersed thereafter, the P-38 delivered a concentrated stream of fire regardless of range. Johnson and his team were the only prewar designers to fully exploit the notion that powering a single seat fighter with two engines had the advantage of halving the per-engine weight penalty of pilot, armament, instrumentation, and flight controls, though the Lightning's success owed at least as much to Johnson's unorthodox approach as to the inherent advantages of the scheme.[61]

The XP-38 (X for experimental) first flew in January 1939 and proved to have spectacular performance, but mismanagement at Lockheed delayed development. Battle-worthy versions of the P-38 did not enter service until the summer of 1942, over nine months behind schedule.[62] The source of the problem was cash-and-carry orders from Britain for patrol bomber versions of the twin-engine Electra transport, desperately needed to stem the U-boat menace. These produced immediate profits for a cash-starved Lockheed, but stripped the YP-38 (Y for service test) program of first-line engineers, draftsmen and machinists. The resultant delay had the doubly adverse effect of depriving the USAAF of a world-class fighter at the outbreak of hostilities and of souring the Army Air Forces on Lockheed as a contractor, a fact that no doubt played a role in restricting P-38 procurement. Moreover, the AAF initially misused the P-38 as a low-altitude battlefield air superiority aircraft in North Africa where it was outclassed by the Bf 109 and Fw 190, suffering a taint to its reputation that was never completely erased.

The P-38's tactical profile was utterly unlike that of any other World War II fighter: it had a good rate of climb and excellent speed, a high service ceiling, a spectacular zoom climb, a good turn radius, and heavy firepower. Equipped from the start with jettisonable drop tanks for range extension it had by far the longest radius of action of any U.S. fighter until the debut of the P-51. A slow initial roll rate, the product of the mass of its two in-line engines mounted well outboard of the center of gravity handicapped the P-38's air to air capability. These characteristics called for tactics that were quite different from those that worked well with other U.S. fighters. P-38 pilots in the Pacific generally adapted well—getting into a turning dogfight with the lighter Japanese fighters was a critical mistake for *all* U.S. fighters, not just the P-38—but only a minority of pilots in the European theater learned to exploit the P-38's capabilities effectively. Complicating matters, the P-38's size and complexity intimidated many pilots and its high wing loading called for initial climb speeds higher than those to which neophyte pilots were accustomed if loss of an engine on takeoff were to be survived.

In addition, technical problems—all fixable—reduced the P-38's effectiveness over Europe: its cockpit heater was inadequate for winter operations over Germany and its intercoolers, the ducting that cooled the outflow from the turbo superchargers, were *too* efficient, reducing the air/fuel mixture to sludge in frigid, moist winter air at high altitudes. This led to blown engines deep in enemy territory when they were most needed, and twin-engined redundancy had little value in a dog fight. In consequence, the P-38 fought over northern Europe at a serious disadvantage: in 90 days of combat beginning 28 December 1943, the 20th Fighter Group, the most highly decorated P-38 group in the European theater, suffered 54 pilots lost to 52 kills awarded.[63] There is bitter irony in the consideration that if the two groups of P-38s operating over Germany in December of 1943 had been deployed six months earlier, something entirely within the realm of the feasible, they would have been available for both Schweinfurt missions and would have fought in the warm skies of summer and early autumn. The problem was one of vision, not design.

The P-38's slow initial roll rate was partially ameliorated by the provision of hydraulically boosted aileron control. Used aggressively in larger numbers and with appropriate tactics, the P-38 had the potential to have done in mid- to late 1943 what the P-47 did and more. In the event, the P-38 played a significant, albeit subsidiary, role in the defeat of the *Luftwaffe* fighter arm. Under the designation F-5, the P-38 was the USAAF's most important American-produced photoreconnaissance aircraft.

The Illushyn Il 2 Shturmovik

Produced in huge numbers, the heavily armored, single-engine Illushyn Il-2 *Shturmovik* was the most important Soviet ground attack aircraft of the war and inflicted serious losses on the *Wehrmacht* from the summer of 1944 to the end of the war.

A 1938 design powered by a liquid-cooled Mikulin V-12 engine in the 1,700-horsepower range, the result of progressive development of a late 1920s BMW original,[64] the Il-2 entered service in the summer of 1941 and was built for one thing and one thing only: low-altitude ground attack. A straightforward design, the Il-2 incorporated little that was novel beyond the extensive structural use of hardened steel to provide armor protection to the engine, cooling system and crew. It was armed with two unsynchronized wing-mounted 23 mm cannon and, in two seat versions, with a flexible 7.62 mm or 12.7 mm machine gun fired by the rearward-facing observer. The IL-2's bomb load was not particularly impressive: 600 kg (1,321 lb) of bombs and rockets carried externally on underwing mounts—the P-47 carried a heavier weight of ground attack ordnance by some 12 percent. This was more than compensated for by its resistance to battle damage and the numbers in which it was produced, more than 36,000 by war's end.[65]

The Consolidated B-24

The strategic importance of the Consolidated B-24 Liberator derives in the first instance from its success as an anti-submarine patrol aircraft in the Battle of the Atlantic and in the second instance from the immense amount of resources the B-24 program absorbed, both in absolute terms and relative to the aircraft's operational effectiveness as a bomber. The B-24 had many technical and tactical deficiencies, but possessed one critically important virtue, range, that was of critical importance in anti-submarine patrols, particularly in combination with a relatively high cruise speed. Small numbers of VLR (for very long range) Liberators were deployed in this role by the British from June of 1941. Progressively upgraded in capabilities, they were procured in modest numbers but to decisive effect; a mere two squadrons of later VLR versions of the Liberators closed the mid-Atlantic "air gap" south of Greenland in the spring of 1943, sounding the death knell of the U-boat force.[66] Early versions of the Liberator, though not battle worthy, were the premier Allied long-range VIP transport and served in this role in small numbers throughout the war. Though substantially less effective as a high-altitude daylight bomber than the B-17, the B-24 made a significant contribution to the Combined Bomber Offensive. From late 1942 until the operational debut of the B-29, the B-24 was the principal USAAF heavy bomber in the Pacific where its superior radius of action conferred important benefits and where its vulnerability to battle damage could be tolerated.

Produced in larger numbers than any other U.S. aircraft of World War II—over 18,000 were built[67]—the B-24 was designed to a March 1939 contract and first flew that December, but was not fielded as a battle-worthy bomber until mid-1942. Procured by the USAAF to supplement the B-17 with a more modern aircraft, the B-24 was powered by four Pratt and Whitney R-1830s, a hedge against demands on production of the Wright R-1820. A large part of the B-24's appeal lay in the supposed efficiency of its wing's Davis airfoil. Designed by California entrepreneur David Davis in a pseudoscientific process of inspired guesswork, the airfoil did not differ appreciably in performance from similar NACA airfoils. That, however, was not appreciated at the time, and Convair's decision to give the B-24 a wing of unusually high aspect ratio, that is a wing that was long as a function of its span, resulted in a wing that was uncommonly efficient at low and medium altitudes and was largely responsible for the aircraft's superior range.[68] The B-24 proved amenable to modifications that enhanced its effectiveness in the anti-submarine role, notably radar installations and a variety of forward-firing ordnance, ranging from 20 mm cannon to stub wing-mounted, 5-inch high-velocity rockets.

Early models of the B-24 had neither turbo-supercharged engines, self-sealing fuel tanks, nor effective defensive armament, and extensive modifications were required to make the aircraft battle worthy, a threshold reached in mid-1942 with the B-24D.[69] At that point, the B-24 was armed with power operated twin .50 cal. dorsal and tail turrets, but, unlike contemporary B-17s, had no belly turret. Later models were fitted with a retractable belly turret and were as heavily armed as the B-17. Not until the J model of 1944 were problems with the flight controls largely worked out, and problems with the leaky fuel system were never completely resolved.

As noted earlier, the B-24 was inferior to the B-17 as a high-altitude daylight bomber in every tactically meaningful parameter except range. That, however, became apparent only with accumulated combat experience, by which time the USAAF had committed itself to procuring the B-24 in large numbers. That decision in itself produced problems, notably in efforts by the Ford Motor Company to mass-produce the B-24 using automotive production methods at the enormous Willow Run facility built specifically for the purpose. In fact, automotive and aircraft production methods were fundamentally different and by the time Ford engineers had mastered the new medium, the B-24 was approaching obsolescence.

The Messerschmitt Bf 110

The Bf 110's strategic importance lies first and foremost in its failure as a heavy day fighter, a major reason for German defeat in the Battle of Britain. Fitted with air intercept radar, the Bf 110 proved to be an excellent night fighter and exacted a heavy toll on RAF Bomber Command from 1942 on, a matter of lesser, but still considerable, strategic importance. In addition, rocket-armed Bf 110s were effective against American heavy bombers during the USAAF unescorted daylight precision bombardment campaign of 1943.

Designed to a 1934 specification calling for a long-range heavy fighter, the twin-engine Bf 110 first flew in 1936, powered by two DB 600s. The DB 601 engine and, later, the DB 605 powered the operational versions. Fast for the time, well armed, and with excellent handling characteristics, the Bf 110 was handicapped by the fact that it was a multi-place aircraft with provisions for a gunner and radio operator, provisions that meant greater size and weight. In consequence, the Bf 110 had insufficient maneuverability to survive in combat against single-engine fighters although this did not become apparent until the Battle of Britain. The Bf 110 turned out to be a superior night fighter since it had sufficient power and payload capacity to carry the requisite air intercept radar and specialized receivers, a radar operator, and heavy armament. Of equal importance, it was an excellent instrument platform. The provision of upward-firing fuselage-mounted 20 mm cannon from August 1943 gave Bf 110 night fighters an unprecedented lethality that the RAF recognized only belatedly. That the installation was developed in the field by an enlisted armorer speaks volumes for low level German initiative and the basic soundness of the Bf 110's construction.[70]

The same power and payload capacity that made the Bf 110 a superior night fighter supported the provision of two under-wing launchers for 210 mm bombardment rockets for daylight use against USAAF heavy bomber formations. This was the *Luftwaffe*'s only air-to-air weapon that outranged the American bombers' .50-caliber defensive armament and the only one capable of degrading the integrity of B-17 defensive formations. Brutally effective when employed in combination with single-engined fighters, the rocket-armed Bf 110 enjoyed a brief heyday before it was put out of business by drop tank-equipped U.S. fighters.

The Boeing B-29

Boeing B-29s reduced the major Japanese cities to ashes in firebombing attacks from March of 1945 and dropped the two nuclear bombs that ended the war. Also of considerable strategic significance, the B-29 program absorbed immense quantities of resources, out-spending the Manhattan Project that produced the A Bomb by some $3.75 billion to $2 billion in 1945 dollars.[71] Finally, B-29s played a major role in cutting off Japanese maritime commerce by means of aerial mines.

Power plants aside—a point that is debatable—the B-29 was far and away the most technologically advanced production aircraft of World War II. Indeed, save for its piston engines and unswept wings, it had more in common with the jet bombers of the 1960s than with contemporary designs. In addition to an unprecedented radius of action in excess of 2,000 miles and a service ceiling above 30,000 feet, the B-29 was the first mass-produced operational bomber with a fully pressurized crew compartment and effective, remotely controlled, defensive armament. The power plants, four R-3350 four-row radial engines, each with two turbo superchargers, were an impressive technological achievement in their own right, and posed major developmental problems stemming partly from sheer complexity and partly from the altitudes at which they had to operate. The B-29 was the USAAF's second most accurate bombing platform next to the B-17.[72] Ironically, Hiroshima and Nagasaki aside, it had its greatest strategic impact dropping incendiaries at low altitudes, a mission where many of its advanced design features were irrelevant.

The Junker Ju 52

A handful of Ju 52 transports forwarded to Spain by Hitler in the early days of the 1936–39 Spanish Civil War provided the critical increment of support that prevented the collapse of the Nationalist cause by ferrying elements of the regular Army of Africa from Morocco to Seville. Although the subsequent drive on Madrid stalled in November 1936, the Nationalists ultimately prevailed, keeping Spain out of World War II and placing the diplomatic tenacity of Spanish dictator Francisco Franco between Hitler's armies and Gibraltar. In addition, German paratroops dropped from Ju 52s or descending in gliders towed by them played a major role in the May 1940 conquest of the Low Countries and the preeminent role in the April 1941 capture of Crete. Finally, the success of the *Luftwaffe* transport arm in supplying the Demjansk pocket, cut off by the Soviet 1941–42 winter offensive, encouraged *Luftwaffe* chief Hermann Goering to believe that he could similarly supply Stalingrad the following winter. Perversely, the Ju 52's successes, all functions of its sound design, led Goering and the Third Reich into disaster.

A late 1920s design with corrugated aluminum skin, three engines, and fixed landing gear, the Ju 52 had modest performance in comparison with its principal Allied counterpart, the C-47. Its maximum and cruising speeds were 168 mph and 124 mph to the C-47's 185 mph and 230 mph and it carried a payload amounting to about 38 percent of its maximum takeoff weight in comparison to the C-47's 45 percent.[73] On the plus side, it was exceptionally rugged, possessed excellent handling characteristics, and was easily maintained in the field under the most primitive of conditions, virtues that kept it in production well after the termination of hostilities in 1945. Among mass-produced World War II transports, it was second only to the C-47 as a parachute deployment aircraft and was capable of handling surprisingly bulky loads.

Conclusions

What conclusions can we draw from the preceding analysis? First, our exercise in rank-ordering supports and gives substance to the proposition that aircraft design *was* a major driving factor in the conduct and outcome of World War II. How many aircraft a country produced was important, but which aircraft it fielded and how well suited they were to their respective nations' grand strategies was crucial. Timing was also a critical factor, which is another way of saying that those establishments which accurately anticipated their strategic needs and allocated their resources accordingly were able to deploy the aircraft they needed in time to have strategic effect. Numbers were important, but it had to be the right aircraft at the right time. To be sure, talented designers were essential to the process, but as we have seen the availability of first-rate aero engines was more likely to be the limiting factor than airframe design talent, a point we shall expand on below. To anticipate another point to be expanded upon, Nazi Germany and the Imperial Japanese Navy did a brilliant job of anticipating their short-term aircraft needs, taking full advantage of the fact that the strategic initiative was theirs to begin with. By contrast, Britain and the United States, more precisely the RAF, USAAF, and U.S. Navy, did a far superior job of planning and preparing for a long war. That is particularly impressive in light of the fact that, as our exercise makes clear, the design of aircraft for strategic applications was inherently more demanding than that for tactical applications and much of the prewar RAF and USAAF effort was directed toward strategic applications, both defensively and offensively on the part of the RAF and offensively by the USAAF.

At a lower level of abstraction, a breakdown of aircraft by nationality underlines just how important American air power was to World War II and how much of that importance was directly dependent on the quality of power plant and airframe design.

It also underlines the importance of America's massive commitment of engineering skill and economic resources to the air war. No less than ten of the twenty-one aircraft on the list are American, followed by four British, four German, two Soviet, and one Japanese. Note, however, that six of the ten American aircraft are in the bottom half of the list and the seventh, the P-47, is at the middle, an accurate reflection of the time needed to mobilize American resources, intellectual as well as productive. Conversely, the fact that the top two aircraft on the list are German provides eloquent testimony to the operational benefits accruing to an aggressor who attacks at the time and place of his choosing. That the second, the Ju 87 Stuka, was operationally effective *only* at the time and place of the aggressor's choosing further underlines the point. That the other two German aircraft on the list are near the bottom, and that both make the list largely because of their contribution to strategic failure, testifies to the incoherence of Nazi strategy and resource allocation when applied to a prolonged war. The location of the only Japanese aircraft on the list, the A6M Zero, makes the same point with regard to Japan. By contrast, the fact that three of the four British aircraft are in the top third of the list provides eloquent testimony to the remarkable prescience of those responsible for Britain's technical preparations for war in the air. So, too, does the fact that one of the American aircraft, the P-51 Mustang, was designed in response to a British requirement and achieved tactical success and strategic importance by virtue its British engine.

In functional terms, the list includes two transports, one specialized patrol aircraft and eighteen combat aircraft. Of the eighteen combat aircraft, no less than eleven were designed as day fighters, including three of the top four and eight of the top eleven, although one of the eighteen, the Bf 110, makes the list primarily because it failed in its primary role. The preeminence of day fighters confirms conventional wisdom concerning the importance of air superiority, both as perceived before the war and as things played out, but with an interesting twist: the importance of long-range fighters was almost entirely unanticipated during the interwar years and only the *Luftwaffe* entered the war with an aircraft, the Bf 110, designed for that role. From the standpoint of those who wrote the specifications, the long-range capabilities of the P-38, P-47, and P-51 were entirely serendipitous. From the standpoint of those who designed them, they were anything but, if only because they took full advantage of their remarkable engineering skills and the impressive power plants available to them to build in a great deal of payload reserve.

Interestingly, four of the remaining eight combat aircraft are four-engine bombers, B-17, Lancaster, B-24, and B-29, although one of the four, the B-24, was included in large part because of its success in an ancillary role. That, too, is more or less in accordance with conventional wisdom, at least as promulgated by the USAAF and RAF. Three of the remaining four combat aircraft are single-engine attack aircraft, the Ju 87 Stuka, SBD Dauntless, and Il 2 Shturmovik, all designed to tight specifications written with specific mission requirements in mind and all outstandingly successful in terms of those specifications, and not much else. No surprises there.

What *is* surprising is the relative absence of twin-engine combat aircraft. The only two to make the list by virtue of operational success, the P-38 and the Mosquito, were remarkably radical and uncommonly successful designs. Particularly striking is the absence of twin-engine bombers. One reason for this phenomenon lies in the fact that certain essential items of equipment, notably defensive armament installations and the gunners who manned them, were of a fixed size and weight and could not be scaled down. Each such installation thus comprised a greater proportion of the gross weight of a smaller aircraft than a larger one. Every bomber on the list except for the Mosquito was defended by powered gun turrets, and turrets could only be made so small, so the same point applies to drag as well.[74] Another reason is that structural materials, notably the rolled aluminum sheet of which most aircraft on the list were constructed,[75] were made to a standard thickness. Although the point needs to be investigated more thoroughly, the skin on a four-engine bomber would thus be thinner and lighter relative to the total weight of the aircraft than that of a twin-engine bomber. To exploit these realities in the design process clearly posed major engineering challenges—only Britain and America fielded operationally successful four-engine bombers—but the payoff in greater range, bomb load, service ceiling, or some combination thereof clearly had great strategic importance. That is not to say that twin-engine bombers were unimportant strategically. Rather, it is to say that such aircraft had substantially more modest operational capabilities, and thus less strategic importance. They were also more or less interchangeable. Light and medium bombers performed useful work, but they did not and could not carry the fight to the enemy as did their larger brethren.

Progressing from the general to the specific, engines were the critical limiting factor in aircraft design where extreme performance was required. With the sole exception of the Zero, every aircraft on the list with outstanding high-altitude performance and/or with exceptional range and payload characteristics was powered by an exceptionally capable engine. In concrete terms, the aircraft in question were powered by the Daimler Benz 601 or Daimler Benz 605 (Bf 109); by the Rolls-Royce Merlin (Hurricane, Spitfire, Lancaster, Mosquito, and P-51); or by turbo-supercharged American engines (B-17, P-47, P-38, B-24, and B-29). To this short list, we can add the Rolls-Royce Griffon-powered Spitfire PR XVI, mentioned earlier. Going beyond our list, mediocre aircraft powered by first-class engines were numerous: to cite two prominent examples, the Handley Page Halifax and some versions of the Curtiss P-40 were powered by Merlins. With the sole exception of the Zero, the converse was not true: insofar as combat aircraft were concerned, mediocre engines powered mediocre aircraft.

Moreover, the strategic importance of high-performance aero engines was magnified by the time required for their development—a minimum of three years by the beginning of World War II[76]—and it is in this context that the remarkable farsightedness of British preparations for war becomes manifest. Government subsidies for the development of high-performance aero engines with military potential were maintained throughout the 1920s and 1930s, although ironically development of the Rolls-Royce "R" was financed with a private contribution of £100,000.[77] Nor was Rolls

Royce, Britain's only producer of high-performance aero engines. Bristol developed and fielded a family of high-performance, air-cooled radial engines that powered some versions of the Lancaster and Halifax, and by 1944 could have provided a capable substitute for Rolls-Royce engines in fighter applications had one been needed.[78] In addition, Napier developed the liquid-cooled 24-cylinder "H" Sabre, the most powerful aero engine to see operational service until the debut of the Wright R-3350, and if the strategic impact of the Sabre-powered Hawker Typhoon and Tempest was comparatively modest, it was not because of the inadequacies of their power plants.[79] Finally, Rolls Royce developed a more powerful successor to the Merlin, the Griffon, in time to see operational service in later versions of the Spitfire. As noted earlier, it was insurance that was not needed.

The American achievement was equally farsighted and more innovative, much of it the product of Army Air Corps initiatives. Considering the United States safe behind its ocean frontiers, Congress stopped subsidizing the development of high-performance aero engines with the onset of the Great Depression, forcing the Army and Navy to depend on engines developed for the civilian market.[80] That meant air-cooled radial engines designed for maximum cruise efficiency at low and medium altitudes, a path down which the Navy had already started as a result of the air-cooled engines' superior power to weight ratios, reliability, and ease of maintenance. For likely Navy missions, attack and defense of ships and maritime patrol, none of them requiring a particularly high service ceiling, that made perfect sense. But Army airmen, looking ahead to a European war, with their utter unpreparedness for World War I clearly in mind and with an eye on developments abroad, saw the need for higher speeds and service ceilings than any conceivable civilian requirement would demand. As already related, their response was on two fronts: contracting with General Electric to develop the turbo supercharger and working aggressively in cooperation with the oil industry to develop aviation gasolines that would support higher compression ratios without pre-ignition "knock." Despite strong resistance on the part of the Army Staff—high-octane gasoline was considerably more expensive than that in common use—the Air Service prevailed, and by the 1930s had managed to obtain supplies of 100-octane aviation gasoline and engines modified to take full advantage of the higher compression ratios it permitted. Integral to this achievement was success in convincing the airlines that greater speed meant more profits, thus providing a civilian market for high-octane fuels, a market ensured by the surprisingly robust health of the American aviation industry during the Depression.[81] A final factor in the triumph of 100-octane aviation gasoline was America's position as a major producer of crude oil, for the reason that the production of a barrel of 100-octane fuel required many more barrels of crude oil than the production of a barrel of 87-octane gasoline, the standard aviation fuel for the oil-starved Third Reich and Japanese empire.

Meanwhile, to hedge its bets, the Army, understanding that high-performance liquid-cooled engines offered higher maximum speeds in fighter applications given the state of the art in the mid-1930s, was able to provide the Allison division of General Motors with a modest subsidy to develop its twelve-cylinder V-1710. For reasons already addressed, the V-1710 never achieved its full potential except in the P-38 and then only when turbo supercharged.

In the greater scheme of things, the V-1710 was a minor chord in a great symphony. The major chords were struck by 100-octane aviation gas and the turbo supercharger. Not least among the Army's achievements lay in apprising the RAF of the benefits of 100-octane gasoline, benefits that perfectly fit the tactical demands that would fall on Fighter Command's Hurricanes and Spitfires in the Battle of Britain. The responsible British authorities saw their opportunity clearly, arranged for the appropriate modifications to their engines, and 100-octane fuel played large in the Battle of Britain. It was to play large in the subsequent successes of the Allied air forces in general.

As for the turbo supercharger, the high-altitude skies above Western Europe were the critical theater of the air war, and until the P-51's operational debut at the very end of 1943, the only U.S. aircraft capable of engaging in combat there were powered by turbo-supercharged engines. Simply put, without the turbo supercharger, the application of America's aerial might against the Allies' most dangerous opponent would have been delayed by at least a year. Good as it was at low and medium altitudes, The F4F was outclassed in service ceiling and high-altitude speed by contemporary versions of the Bf 109 and Fw 190. Even the second generation of U.S. Navy fighters that entered service from the late summer of 1943, the Grumman F6F Hellcat and the Vought F4U Corsair, would have been outclassed at high altitudes by later versions of the Bf 109 and Fw 190 and at low altitudes would have fought at little better than a par.[82] More crucially, the United States would have had no bomber with the service ceiling needed to survive in daylight over Europe. Would the USAAF and American industry have found other avenues to high-altitude performance? Perhaps, though it is difficult to see how they could have been deployed quickly or that they would have worked as well.

What lessons can we distill from our exercise? First and most basic, technological competence and strategic vision are essential, but only if applied in conjunction with one another. The Allies won World War II in the air largely because the leaders of the RAF, the USAAF, and the U.S. Navy's aviation establishment had a clear strategic vision and a good sense of what was feasible in terms of the aircraft performance needed to make their respective visions reality. That their strategic visions were imperfect in important particulars was inevitable. What was important was that they *did* have a vision and that it drove technology in positive directions. Technological cleverness in isolation is not enough, as the *Luftwaffe* and Nazi Germany's aviation industry clearly demonstrated. Indeed, it can be counter-productive.

In the case of World War II, the combination of strategic vision with technological competence was rendered particularly important by the considerable amount of time needed to develop certain critical technologies. One hundred-octane aviation gas and the turbo supercharger are salient examples with British development of high-performance aero engines not far behind. The development of high-performance engines in the United States is a special case since they were developed for civilian applications, but it only adds to the luster of the Army Air Forces planners that they were able not only to effectively harness a civilian technology to their needs, but advance the civilian technology in the process.

That having been said, it is important to remember that our exercise underlined the considerable advantages accruing to the strategic aggressor. In being able to determine the time and place of their attack, and in obtaining aircraft suitable for their chosen operational methods, the Third Reich and Imperial Japan came closer to success than we commonly admit today or was comfortable then. The fact that it is easier to design for tactical than for strategic advantage—and this applies to technologies other than aircraft—renders ideologically motivated aggressors all the more dangerous however unbalanced their world views may be. In that there is surely a lesson for today.

And what of aircraft design proper? It was, as I hope I have demonstrated, strategically pivotal, though in the final aggregate not as important as the underlying factors just enumerated. It is worth noting in this respect that much of America's success was attributable to the sheer depth, breadth, and vitality of America's aviation industry. In hindsight, it is easy to see that in the aftermath of World War I looking forward to what most competent military professionals saw as an upcoming global conflict, aircraft technology was a critical variable, perhaps *the* critical variable. It was not so apparent at the time and air power advocates in Great Britain and the United States had a difficult time selling their case. Today, to many aviation technology does not seem to have the same strategic primacy that it did in the interwar period, but we say this with full wisdom of hindsight and are defining aviation technology narrowly in the process. Broadly defined to include information gathering and electronic warfare capabilities across the entire aerospace spectrum, it surely retains its importance and demands the same clear vision for the future in the aftermath of the Cold War that it did in the aftermath of World War I.

Notes

1. In contrast, air power made major contributions to the defeat of the U-boat menace in the Battle of the Atlantic. Note, too, that the destruction of Japanese shipping was powerfully assisted in the final stages by B-29–dropped mines and that carrier aviation made significant contributions as well, notably in sinking eight of the twelve large rail ferries that connected Hokkaido to Honshu and rendering the other four unusable. See Richard B. Frank, *Downfall: The End of the Imperial Japanese Empire* (New York: Penguin, 2001), 154–59. That having been said, submarine attacks were well on the way to reducing the size of the Japanese merchant marine below that needed to sustain the war economy before aerial mining or carrier aviation became significant factors.

2. Of the examples cited, all except Stalingrad are unequivocal. The *Luftwaffe* controlled the skies above Stalingrad during the early stages of the battle, but Soviet strength in the air grew apace and by the time the Soviets launched their November 1942 counterattack, the Red Air Force had achieved rough parity. Although the Red Air Force lacked the strength to drive the *Luftwaffe* from its bases, neither was the *Luftwaffe* able to contain the depredations of marauding Soviet fighters which wrought havoc on German aerial resupply operations with decisive effect.

By German admission, over 320 *Luftwaffe* transports were shot down, representing a major loss in trained aircrew that had strategic implications well beyond the battle's outcome. Von Hardesty, *Red Phoenix. The Rise of Soviet Air Power, 1941 1945* (Washington, D.C.: Smithsonian Institution Press, 1991), 91–120, esp. 110.

3. The only daylight actions among these were the 13 December 1939 Battle of the River Plate in which the German pocket battleship *Graf Spee* was defeated by British cruisers *Ajax, Achilles,* and *Exeter;* the 13 April 1940 Second Battle of Narvik, in which a British destroyer force led by the battleship *Warspite* destroyed a German destroyer flotilla; and the inconclusive engagement on 26 March 1943 between U.S. and Japanese cruiser forces off the Komandorski Islands in the Bering Sea.

4. The most important of these were the 8–9 August 1942 Battle of Savo Island, a Japanese tactical victory; the 11 October Battle of Cape Esperance, an American tactical victory; and the two-phase Battle of Guadalcanal, 13 November (a Japanese tactical victory) and 14–15 November (an American tactical and strategic victory); and, finally, the 30 November Battle of Tassafaronga, a Japanese tactical victory, the last of the war.

5. As the Solomons campaign progressed, Allied aerial reconnaissance became progressively more effective and ultimately constituted a major operational advantage. That having been said, particularly in the early stages of the campaign, Japanese skill in avoiding or deceiving allied aerial reconnaissance yielded major tactical advantages, most notably in the Battle of Savo Island, where catapult-launched Japanese spotting aircraft rendered useful service by illuminating the American and Australian battle line with parachute flares.

6. Geoffrey Bennett, *Naval Battles of World War II* (New York: David McKay, 1975), 161–66.

7. See Martin Middlebrook, *The Berlin Raids: Bomber Command, Winter 1943–44* (London: Cassell, 1988), 6–9, for a useful overview of objectives and forces committed.

8. See Middlebrook, *Berlin Raids,* App. 3, Bomber Command Statistics, for British losses.

9. James Campbell, *The Bombing of Nuremberg* (New York: Doubleday, 1974), 142–45. Seven hundred and ninety-five Lancasters and Halifaxes were sent against Nuremberg, of which 94 failed to return, a loss rate of 11.8 percent. Of the total, 14 were downed by FLAK and 2 lost to a midair over the target.

10. *Ranger,* older and smaller than the rest, was deemed unsuited for fleet operations. *Saratoga* was in dry dock undergoing repair for torpedo damage.

11. This is no doubt due to the British reading public's seemingly inexhaustible appetite for accounts of the Battle of Britain and to the historical bent of the British technical aviation press. Interestingly, however, the best general account of the Battle of Britain in my view, and one of the few air campaign histories to systematically and effectively connect aircraft design to strategic consequences, is by a novelist: Len Deighton, *Fighter: The True Story of the Battle of Britain* (London: Jonathan Cape, 1977). T. C. G. James, *The Battle of Britain,* ed. Sebastian Cox (London: Cass, 2000), the official RAF account written in 1943–44, is the best and most complete historical narrative, but says little about aircraft design and was published only in 2000.

12. This point is effectively made by attritional data presented in Williamson Murray, *Strategy for Defeat: The Luftwaffe, 1933–1945* (Maxwell Air Force Base, Ala.: Air University Press, January 1983), esp. Table XXX, "German Losses by Theater, January–November 1943," and Table XXI, "German Fighter Losses, 1943 (Number of Aircraft)," 148–49.

13. Service ceiling is the highest altitude at which an aircraft can sustain a rate of climb of 100 feet per minute; John D. Anderson, Jr., *Introduction to Flight,* 2d ed. (New York: McGraw-Hill, 1985), 285–88. The B-17G had a service ceiling of 35,000 feet and the B-24J a service ceiling of 28,000 feet; William Green, *Famous Bombers of the Second World War* (Garden City, N.Y.: Hanover House, 1959). While these figures apply to individual aircraft, USAAF heavy bombers penetrated in massed formations. Formation flying involves considerable jockeying around and formation performance is dictated by the poorest performing aircraft in the formation. In consequence, actual penetration altitudes were typically in the 25,000–27,000 foot range for B-

17s, although occasionally as high as 29,000 feet, and around 22,000–23,000 for B-24s. These figures in note 15 are based on examination of the records of the 2nd Bombardment Division, cited below, a B-24 unit; and the 95th Bombardment Group (Heavy), a B-17 unit, for example, the records of 95 BG Mission #41, the 14 October 1943 Schweinfurt raid, National Archives II, College Park, Md., RG 18, Stack Area 190, Row 58, Compartments 4-5, Shelves 7-3, Box 333 (hereafter cited in the format NAII/RG18/190/58/4-5/7-3/333).

14. The B-24's notoriously leak-prone fuel system was a particular problem; structural weakness was a factor as well.

15. Headquarters 2nd Bombardment Division, "Summary of Mission Number 138" and "Minutes of Critique, Mission 14 October 1943," NAII/RG18/190/59/16/5-6/2621.

16. The Bf 109 was designed by Willy Messerschmitt in his capacity as chief designer of the *Bayerische Fleugzeugwerke* (Bavarian Aircraft Works), from whence the designation Bf. The name of the company was changed to Messerschmitt AG (*Messerschmitt Aktiengesellschaft*) in July 1938 and designs subsequent to the Bf 110 received the Me prefix; J. R. Smith and Antony Kay, *German Aircraft of the Second World War* (London: Putnam, 1972), 472.

17. Dénes Bernád, *Heinkel He 112,* Aircraft Number 159 (Carrolton, Tex.: Squadron/Signal Publications, 1996), 4–5.

18. William Green, *Famous Fighters of the Second World War* (London: McDonald, 1957), 16; some 36,000 *Shturmoviks* were produced, see note 57, below.

19. L. J. K. Setright, *The Power to Fly: The Development of the Piston Engine in Aviation* (London: Allen & Unwin, 1971), 101–2. So efficient was the design of the Merlin's centrifugal flow compressor that it was incorporated into the Whittle turbojet, progenitor of the Rolls-Royce Nene of the late 1940s. Produced under license in the Soviet Union, Nenes powered the MiG 15 and later versions powered the MiG 17, still a competitive air-to-air fighter in the 1970s, eloquent testimony to the efficiency of the Merlin's compressor!

20. The genesis of both the Hurricane and Spitfire lay in a 1930 specification for an eight-gun fighter issued by the Royal Aircraft Establishment (RAE), the section of the Air Ministry responsible for aircraft requirements and specifications. The RAE issued a further specification for a Hawker experimental fighter in October 1934 with no further stipulations. The performance specification that led directly to the Hurricane and Spitfire was issued in April 1935; Colin Sinnott, *The Royal Air Force and Aircraft Design, 1923–1939* (London: Cass, 2001), 77, 86–87. Inasmuch as the RAE's technical expertise and operational authority was resident in its senior RAF members and worked closely with the Air Staff, I have referred to its specifications as RAF specifications for simplicity.

21. T. C. G. James, *The Battle of Britain, RAF Official Histories,* ed. Sebastian Cox (London: Cass, 2000), 332, 338, 368, App. 2, 5, 14, and 18, addressing Fighter Command sector organization and order of battle on 7 July, 8 August, 7 September, and 30 September 1940. Eight Hurricane squadrons were added between 8 August and 7 September, including one Canadian, one Czech, and two Polish squadrons.

22. Owen Thetford, *Aircraft of the Royal Air Force, 1918–1957* (London: ??, 1957), 154–58.

23. Information to the author from John D. Anderson, Smithsonian Institution, spring 2002.

24. Smith and Kay, *German Aircraft,* 486.

25. Victor Bingham, *Major Piston Aero Engines of World War II* (London: Airlife, 1998), 115.

26. The R-1820 was originally designed to burn aviation gasoline with octane ratings in the mid- to high 80s. The Air Corps persuaded Curtiss Wright to produce a version for the Martin B-10 designed to operate at higher compression ratios, exploiting the greater resistance to "knock" provided by 100-octane gasoline. The twin-engine B-10, which entered service in 1932, proved to be substantially faster than existing bombers—and fighters—demonstrating the value of high-octane gas and engines designed for it. A significant problem was the greater cost of high octane gasoline, a problem solved by persuading the airlines that the increased speeds that the new engine-fuel combinations made possible would more than pay for the more costly fuel. A commercial market for high-octane aviation gasoline ensured a ready supply for the military. The Army

Air Corps' technical development branch was the driving spirit behind these developments, powerfully assisted by reserve Air Corps Captain James Doolittle acting in his capacity as a Shell Oil Corporation executive.

27. By the late 1930s, all but the smallest aero engines had integral mechanical superchargers to atomize the fuel charge and to provide a modest increase in boost. My reference here is the provision of additional supercharging to increase power at high altitudes.

28. See Daniel D. Whitney, "Getting High in the Sky," *Skyways* 38 (April 1996), for a competent summary of the General Electric turbo supercharger's development.

29. See note 13, above.

30. The B-24 was powered by four turbo-supercharged Pratt and Whitney R-1830s. The power output of the nine-cylinder R-1820 and the twin-row fourteen-cylinder R-1830 varied slightly from model to model and increased somewhat over the course of the war, but was similar from start to finish. Both were excellent engines. The R-1830 had a smaller external diameter and in a well-designed installation produced less drag; the R-1820 had a better power to weight ratio.

31. Ian Hogg and John Weeks, *Military Small Arms of the Twentieth Century* (Chicago: Follett, 1973), 5–72; David R. Mets, *Nonnuclear Aircraft Armament: The Evolution of Aircraft Guns (1912–1945)* (Eglin AFB, Fla.: Office of History, Armament Division, Air Force Systems Command, 1987), 60–68.

32. The ballistic coefficient is determined by the relationship between a projectile's mass and its aerodynamic drag and is generally proportional to the diameter squared. Since mass increases with the diameter cubed whereas drag increases with the diameter squared, effective range increases as a function of bore diameter given the same muzzle velocities, as was the case here. In addition, the U.S. .50 caliber round had a "boat-tailed" bullet with exceptional aerodynamic qualities.

33. *AAF Bombing Accuracy: Continental [and] Overseas, Report No. 1* and *Report No. 2* (declassified from SECRET), not further identified as to origin and undated, though exploiting data from January through December 1944, NAII, RG 18, 190/61/24/2-5, Entry 10. Based on analysis of a mass of training and combat data, this study is analytically sound and mathematically sophisticated.

34. Ibid., *Report No. 2*.

35. Brereton Greenhaus, Stephen J. Harris, William C. Johnson, and William G. P. Rawling, *The Official History of the Royal Canadian Air Force*, vol. 3 *The Crucible of War, 1939–1945* (Toronto: University of Toronto Press with the Ministry of Supply and Services Canada, 1994), 754.

36. Sinnott, *Aircraft Design,* 142, 158–59.

37. In fact, the Vulture was based on the earlier Rolls-Royce V-12 Kestrel engine, but with the blocks re-bored to yield the same cylinder diameter as the Merlin; Victor Bingham, *Major Piston Aero Engines of World War II* (Shrewsbury, England: Airlife, 1998), 134–35.

38. The principal Axis examples were the German Junkers Jumo 222 and Daimler Benz 604, both twenty-four cylinder X engines; and the DB 606, consisting of two DB 601s mounted side by side and driving a single propeller through a common reduction gear. The DB 606 was the only one of these to achieve operation status, powering the He 177 heavy bomber, but was unreliable and prone to engine fires. The only successful World War II liquid-cooled aero engine with more than twelve cylinders was the twenty-four cylinder Napier Sabre that powered the Hawker Typhoon and Hawker Tempest.

39. Anthony Furse, *Wilfred Freeman: The Genius Behind Allied Survival and Air Supremacy, 1939 to 1945* (Staplehurst, Kent: Spellmount, 2,000), 143. Bingham, *Major Piston Aero Engines,* 136–39.

40. Takashi Nishiyama, "Aeronautical Technology for Pilot Safety: Reexamining Deck-landing Aircraft in Great Britain, Japan, and the United States," *Historia Scientiarum* 13 (2003), esp. 16.

41. The empty weight of the A6M2 Zero was less than 4,000 pounds to 6,500 pounds for the Spitfire IX, the most capable version of the Spitfire in operational service in 1943. Both were powered by engines of about 1,300 horsepower. See Pierre Closterman, *Flames in the Sky* (London: Chatto & Windus, 1956), 49–58, for a clinical evaluation by a top-scoring Allied World War II ace of the reasons for the Zero's effectiveness.

42. Nishiyama Takashi, *Technology for Flight Safety in Japan: The Case of the Mitsubishi Zero Fighter* (Unpublished M.A. thesis, Ohio State University, 1995).

43. William Green, *War Planes of the Second World War, Fighters* vol. 4 (Garden City, N.Y.: Doubleday, 1961), 102–3.

44. Ironically, the only exception, and that a partial one, was the Imperial Japanese naval air arm; John B. Lundstrom, *The First Team: Pacific Naval Air Combat from Pearl Harbor to Midway* (Annapolis, Md.: Naval Institute Press, 1984), 486–89. Japanese Navy fighter pilots were taught off-angle deflection shooting, but were unable to take full advantage of their training in fighter-versus-fighter combat because of poor downward visibility over the nose of the A6M Zero.

45. See ibid., 458–68, 477–85, for a comprehensive discussion of U.S. Navy gunnery and tactics.

46. In principle, the drag created by aerodynamic interference at the junction of wing and fuselage is minimized by mounting the wing near the mid-section of a fuselage of circular or elliptical cross-section. This, however, required longer landing gear to ensure propeller clearance during takeoff and landing, and longer landing gear entailed greater weight, greater mechanical complexity, or both. Kartveli solved the problem with telescoping main landing gear struts that shortened before folding upward into the wing. To the best of my knowledge, this was the only telescoping landing gear of World War II to be essentially trouble free, a tribute to Republic's second-tier engineering.

47. John F. Kries, ed., *Piercing the Fog: Intelligence and Army Air Forces Operations in World War II* (Bolling AFB, Washington, D.C.: Air Force History and Museums Program, 1996), 190–213.

48. The principal RAF electronic target-marking aids, GEE (from early 1942) and Oboe (from early 1943), established a bomber's location by reference to ground-based radio transmissions. Their maximum range was thus limited by the curvature of the earth and the Mosquito's greater service ceiling, 39,000 feet to 24,500 feet for the Lancaster, meant that it could receive the guidance signals considerably further from the transmitter, thus extending Bomber Command's reach into Germany; I. C. B. Dear and M. R. D. Foot, "Electronic navigation systems," *The Oxford Companion to World War II* (Oxford and New York: Oxford University Press, 1995).

49. Sinnott, *Aircraft Design,* 203–8.

50. Thetford, *Aircraft of the Royal Air Force,* 156. I am indebted to Lt. Col. Robert Ehler, USAF, for pointing out to me the superior performance of the Mosquito PR XIV and its strategic importance.

51. Ibid., 130–31.

52. Ironically, the best Axis tactical transport of the war was the Japanese LD2, a DC 3 built under license by Nakajima and Showa as the Imperial Navy's standard land-based transport; R. J. Francillon, *Japanese Aircraft of the Pacific War* (New York: Funk & Wagnalls, 1970), 499–503. Some 485 were produced by Showa and 70 by Nakajima.

53. The designation C-53 was applied to the initial military conversions without the large rear door and provisions for carrying heavy cargo. For convenience and reflecting common usage, I have referred to all military versions as C-47s.

54. Thetford, *Aircraft of the Royal Air Force,* 190–91. The payload figure represents a maximum overload and the radius of action is a conservative estimate based on a stated range of 1,500 miles.

55. Ibid; R. E. G. Davis, *TWA: An Airline and Its Aircraft* (McLean, Va.: Paladwr, 2000), 38.

56. Stephen L. McFarland and Wesley Phillips Newton, *To Command the Sky: The Battle for Air Superiority over Germany, 1942–1944* (Washington, D.C.: Smithsonian Institution Press, 1991), 246–47.

57. Ray Wagner, *Mustang Designer: Edgar Schmued and the P-51* (Washington, D.C.: Smithsonian Institution Press, 1990), 26–29.

58. Thetford, *Aircraft of the Royal Air Force,* 344.

59. Wagner, *Mustang Designer,* 104.

60. The specification called for a maximum speed of 400 mph at 20,000 feet and stipulated that the aircraft be able to climb to 20,000 feet in six minutes; Le Roy Weber, Jr., "The Lockheed XP-38," *Skyways* 69 (January 2004).

61. A partial exception was the British Westland Whirlwind, the only other twin-engine, single-seat fighter to see operational service in World War II. A 1936 design of orthodox configuration, the Whirlwind was powered by two 885-horsepower Rolls-Royce V-12 liquid-cooled Peregrine engines and had an impressive armament of four nose-mounted 20 mm cannon. The Whirlwind first flew in October 1938, but development problems with the Peregrine delayed operational deployment until mid-1941. Faster than contemporary single-engine fighters at low altitudes, the Whirlwind proved to be an effective low-altitude escort and ground-attack fighter, but its operational virtues were not sufficiently compelling to justify large-scale production, the more so as it was the only operational aircraft powered by the Peregrine; Green, *Fighters,* 2:123–25.

62. David W. Ostrowski, "Early P-38 Problems," *Skyways* 40 (October 1996).

63. The 20th Fighter Group Association, *King's Cliffe,* rev. ed. (Moore, Pa.: Sheridan, 2004), 112. This account is particularly valuable for documenting the P-38's strengths and weaknesses from the viewpoint of the men who flew and maintained it. The 20th was dispatched to England in late August 1943 and mounted its first operations in November. The P-38 was available in the European theater in significant numbers only from late December.

64. Bingham, *Major Piston Aero-Engines,* 166–70.

65. Dear and Foot, *Oxford Companion,* "Bombers," 143–50, esp. 146; and "Fighters," 354–63, esp. 359.

66. The first of these were twenty Liberator Is built to British specifications and supplied to the Royal Canadian Air Force for operations from Northern Ireland; Thetford, *Aircraft of the Royal Air Force,* 132–37. These were followed by 139 Liberator IIs from August 1941 and 260 Liberator IIIs, equivalent to a B-24D, from mid-1942; Green, *Famous Bombers,* 85–90; Dear and Foot, *Oxford Companion,* "Air gap, mid-Atlantic," 12–13.

67. Dear and Foot, *Oxford Companion,* "bombers," 146.

68. Walter G. Vincenti, *What Engineers Know and How They Know It: Analytical Studies from Aeronautical History* (Baltimore, Md. and London: Johns Hopkins University Press, 1990), chap. 2, "Design and the Growth of Knowledge: The Davis Wing and the Problem of Airfoil Design, 1908–1945)." The B-24's wing had an aspect ratio of 11.55; that of the B-17 had an aspect ratio of 7.58.

69. Green, *Famous Bombers,* 89–90.

70. Smith and Kay, *German Aircraft,* 501.

71. Kenneth P. Werrell, *Blankets of Fire: U.S. Bombers over Japan During World War II* (Washington, D.C.: Smithsonian Institution Press, 1966), 238, 238n29. A 1945 dollar was worth over twenty times as much as a 2004 dollar.

72. *AAF Bombing Accuracy: Continental [and] Overseas, Report No. 1;* see note 29, above.

73. Smith and Kay, *German Aircraft,* 358, 370; Thetford, *Aircraft of the Royal Air Force,* 190–91. The figures for useful load as a percentage of maximum takeoff weight may be a bit high (my sources do not specify what constitutes empty weight), but should be accurately indicative of the difference in efficiency between the two aircraft.

74. In principle, remotely controlled turrets could be made that were substantially lighter and

produced less drag than their manned equivalents, but this proved to be far more difficult than anticipated. The *Luftwaffe,* RAF, and USAAF all attempted to develop remotely controlled turrets, but only the USAAF succeeded. The first successful such installation was the twin .50 caliber chin turret installed in B-17s from late 1943. The B-29 had the first effective remotely controlled defensive armament system.

75. The exceptions include the Yak fighters and early versions of the Hurricane that made significant use of wood and fabric; the Mosquito with its stressed plywood structure; the Il 2 with important structural elements of hardened steel; and the Ju 52 with corrugated aluminum skin.

76. Setright, *Power to Fly,* 110.

77. Ibid., 97.

78. And in fact, did so in the Hawker Tempest II, a design that entered operational service shortly after the war, powered by a massive four-row Bristol Centaurus radial engine; Thetford, *Aircraft of the Royal Air Force,* 308–9. The Tempest II and its naval version, the Sea Fury, were among the fastest and most capable piston-engined fighters ever built.

79. To be sure, the Sabre was more temperamental than the Merlin, but the main factor behind its limited impact on the war lay in the aerodynamic limitations of the Hawker Typhoon, limited to low altitudes by the its thick wing section. This problem was solved in the Tempest V, the fastest Allied piston-engine fighter of the war, but the Tempest did not enter squadron service until July 1944; Ibid., 306–7.

80. Herschel Smith, *A History of Aircraft Piston Engines* (Manhattan, Kans.: Sunflower University Press, 1981), 70–76, 103–8.

81. America was becoming a motorized society and demand for gasoline engines and fuel remained strong. During 1930–38, average personal income declined by 25 percent in the United States, but the demand for gasoline declined during only two years of that period, and then by only about 5 percent. While only a minority of the demand was for aviation fuel, the benefits of higher octane were appreciated for automotive fuel as well. The maximum octane rating of automobile gasoline increased from 65 to 87 during this period; William David Compton, "Internal-combustion and their Fuel: a Preliminary Exploration of Technological Interplay," *History of Technology,* ed. A. R. Hall and Norman Smith, vol. 7 (London: ??, 1982), 23–36, esp. 34.

82. Eric N. Brown, *Duels in the Sky: World War II and Naval Aircraft in Combat* (Annapolis, Md.: Naval Institute Press, 1988). Written by a highly experienced Royal Navy test pilot who flew the aircraft about which he writes, some in combat, this work is invaluable. Brown's strategic judgments inspire little confidence, but his tactical assessments are as close to definitive as we will ever get.

Rolling Thunder and Linebacker Campaigns: The North Vietnamese View

Merle L. Pribbenow II

North Vietnam has the dubious distinction of having more combat experience against U.S. air power than any other nation in the world. Rolling Thunder, the first U.S. bombing campaign against North Vietnam (1965–68), lasted longer than U.S. air operations in Europe during World War II. When one adds the 1972 Linebacker air campaign against North Vietnam and the almost nine-year bombing campaign against the Ho Chi Minh Trail in Laos, only Iraq, with the air campaigns of Operations Desert Storm and Iraqi Freedom book-ending a twelve-year (1991–2003), low-intensity confrontation against U.S. aircraft over the no-fly zones, faced U.S. air attacks longer. The air battles over Iraq, however, cannot be compared with the battles fought in the skies over North Vietnam. During the course of the war, more than 1,100 U.S. fixed-wing aircraft were lost in combat operations.[1]

The North Vietnamese already had considerable combat experience against the much smaller and less capable French Air Force, and they had no illusions about the threat posed by U.S. air power. During the early 1960s, the North Vietnamese went to considerable pains to conduct their effort to overthrow the U.S.-backed government of South Vietnam in such a way as to avoid provoking an American response—a response that they believed would threaten the survival of the Communist regime in North Vietnam.[2] The Vietnamese Communists, however, viewed air attacks as part of an integrated military effort and anticipated that if the United States "expanded the war into North Vietnam," U.S. air attacks would be accompanied by a ground invasion. The pre-1965 North Vietnamese military preparations for possible U.S. attacks against North Vietnam focused as much on the danger of a ground invasion of North Vietnam as they did on air attacks. The North Vietnamese were so concerned about the threat of an invasion that in 1964 they sought a promise from China that the Chinese would send "volunteer" troops to defend North Vietnam in the event of a U.S. ground invasion.[3]

The first U.S. bombing attack against North Vietnam, conducted on 5 August 1964 after the infamous "Gulf of Tonkin Incident," revealed a number of shortcomings in North Vietnamese air defenses. The North Vietnamese Air Defense–Air Force Service—the Air Defense Command (ADC)—had assumed that U.S. air attacks against the North would be made primarily by medium and heavy bombers flying at medium to high altitudes in World War II–style formations. Vietnamese air defenses were configured to defend against such attacks. The heart of the North Vietnamese defense was their heavy-caliber, slow-firing, high altitude anti-aircraft guns, not the quick-firing low-medium altitude weapons needed to engage fast and highly maneuverable U.S.

tactical aircraft. At the time of the Gulf of Tonkin attack, North Vietnam had forty-five batteries of heavy 85mm, 88mm, 90mm, and 100mm guns but only twenty-seven batteries of lighter 57mm and 37mm guns. The Vietnamese had deployed their guns in widely dispersed positions along probable flight paths rather than close to the actual bombing targets.[4] The U.S. attacks on 5 August 1964 were made by Navy aircraft from aircraft carriers in the Tonkin Gulf. These aircraft flew in at extremely low altitude to make sudden "pop-up" dive-bombing attacks, surprising and confounding the North Vietnamese defenses. Although the gunners managed to shoot down two U.S. aircraft on 5 August, the ADC knew that changes were needed before the next American attack. An official assessment of the ADC's performance during the 5 August attacks reached the following conclusion:

> The Air Defense Service recognized the enemy's character and his strategic intentions, but our assessment of the enemy's tactical plans was not sufficiently specific to enable us to respond to the enemy's actions. We received a report about the enemy's announcement of the bombing, but we were slow to issue the battle alert to our forces and so were taken by surprise. We had not made a detailed study of the characteristics of the enemy's naval aircraft, and in particular we had not studied their attack tactics.[5]

The reference to being "slow to issue the battle alert" after receiving "a report about the enemy's announcement of the bombing" is an oblique admission of the exposure of a serious problem in North Vietnam's ADC structure. Following the alleged second attack by North Vietnamese torpedo boats against American warships on the night of 4 August, President Lyndon B. Johnson ordered the 5 August air strikes and planned a television announcement to the nation, timed to occur just as the bombs were falling to avoid giving the North Vietnamese advance notice of the attack. The air strikes had to be delayed, however, as it took the Navy some time to arm and prepare the aircraft and to move their aircraft carriers into striking position. Back in the United States, half a world and eleven time zones away, President Johnson, who did not want to make his television address after midnight, decided to go ahead with his television address even though he knew that it would be more than an hour before the first bombs would fall. The North Vietnamese General Staff's intelligence office quickly picked up word of the impending air strike over the radio and immediately passed the alert on to the ADC. The entire ADC leadership, however, had been up much the previous night monitoring the "unusual" U.S. activities of U.S. Navy aircraft over the Tonkin Gulf, which they saw as preparatory to an air attack. All ADC units throughout North Vietnam were placed on Condition 1 alert status for a U.S. air strike early the next morning. This meant that all guns, radars, fire control equipment, and command posts were fully manned and battle ready.

By late morning on 5 August, however, when no air strikes materialized, ADC leaders decided that the entire situation had been a "false alarm." They lowered the nationwide alert status to Condition 2 (all units retaining 50 percent of their personnel on ready alert status, with the rest authorized to conduct regular duties or leave their bases for short periods). The entire senior ADC leadership, exhausted and hungry after the long and sleepless night, left the headquarters to return to their homes for the traditional Vietnamese noontime siesta (which, as in most tropical countries, lasted for several hours). When the General Staff notified the ADC Headquarters of President Johnson's speech and ordering the ADC to full battle alert status, the call was taken by the duty officer. Rather than immediately issuing a combat alert, the officer decided that he had to first consult with a senior command officer before issuing the alert order. He had trouble locating the senior officers, however, and by the time the alert order was issued, the attack was already underway. It was a hard lesson for the ADC, and one that they took to heart. In the future, the ADC duty watch always included a senior command-level officer with the power to make immediate decisions on his own.[6]

North Vietnam immediately brought its only air force combat unit, a brand new MiG-17-equipped fighter regiment still in training at an airfield in southern China, back to Vietnam to strengthen North Vietnam's air defenses. It also asked the Soviet Union for surface-to-air missiles, but the request was not approved by the Soviets until the end of 1964, and the missiles did not arrive until April 1965.[7] By the time U.S. bombing of North Vietnam began in February 1965, North Vietnam's anti-aircraft artillery forces had grown only slightly, from twelve to fourteen regiments, as North Vietnam continued to hold out the hope that it could achieve its goals in South Vietnam without triggering a direct U.S. military response.[8] The initiation of sustained U.S. bombing—Operation Rolling Thunder—in the spring of 1965 and the simultaneous introduction of U.S. ground combat units into South Vietnam finally convinced North Vietnam that a direct confrontation with the United States was inevitable. In April 1965, North Vietnam instituted a massive military conscription program and placed its armed forces on a wartime footing. By the end of the year, the North's army had doubled in size, and the size of its air defense forces expanded by 250 percent.[9]

The North Vietnamese were so concerned about the initiation of U.S. bombing that in the spring of 1965 they made a series of desperate requests to their Soviet and Chinese allies for equipment, advisors, and "volunteer pilots" to help to defend North Vietnam. The Soviet Union sent a contingent of several thousand "advisors," and Soviet surface-to-air missile personnel fought in numerous combat engagements against U.S. aircraft. Although the Chinese refused to keep their earlier promise to send "volunteer" pilots to assist the North Vietnamese Air Force, they did send a large contingent of engineer, transportation, and anti-aircraft artillery units (numbering 170,000 soldiers at the highest point, in 1967). These Chinese forces quickly took over responsibility for road construction and air defense duties in the northern tier of provinces along North Vietnam's long border with China. The Chinese presence freed up North Vietnamese forces to concentrate on the areas farther south. During the course of the war, a total of 18 Soviet military personnel and 1,100 Chinese soldiers were killed in action defending North Vietnam against U.S. air attacks.[10]

Soon after Rolling Thunder got underway, the North Vietnamese realized that the Johnson administration had imposed severe restrictions on the bombing of North Vietnam. The U.S. bombing campaign was designed to "send a message" rather than to destroy their regime. Through leaks to the press and the administration's own statements, the North Vietnamese learned that, in their own words, "To carry out his war of destruction against North Vietnam, the U.S. President decided to use the 'escalation' policy proposed by Herman Kahn. This policy was viewed as a political strategy using military resources to gradually put pressure on us. The Americans were certain that 'the more the war expanded, the quicker the war would end.'"[11]

In August 1965, U.S. air attacks expanded to cover most of North Vietnam. At that time the North Vietnamese analyzed the American bombing campaign and found it wanting:

> Even though they continue to "escalate" their attacks, the enemy is displaying several weaknesses. First, the scale and the pattern of the bombing have so far remained unchanged. Second, because of political and diplomatic constraints, the Americans are not yet able to make direct attacks against Hanoi and Haiphong. Their primary objective continues to be to attack our lines of communications in order to block the flow of supplies to South Vietnam.[12]

Buoyed by this conclusion, the North Vietnamese began to disperse the concentration of air defense forces around the Hanoi area (which initially totaled one-third of the entire ADC), sending the units out to protect their supply lines and to gain combat experience.[13] It would be one full year before the ADC finally recalled these units to protect the key cities of Hanoi and Haiphong, and by that time North Vietnam's air defense forces would have grown enormously. The North Vietnamese had been given a valuable respite to learn how to deal with U.S. air tactics and technology. It was a respite they desperately needed.

The North Vietnamese ADC wrestled with a host of problems during the first years of Rolling Thunder. Vietnamese Air Force fighters suffered such heavy losses during 1965 that the Air Force concluded that "if this trend continues we will not be able to continue combat operations over the long term" and held meetings to determine "why the combat effectiveness of the Air Force is so low."[14] When the much more capable MiG-21 was first introduced into the North Vietnamese Air Force's inventory in early 1966, it was completely ineffective, and it took the Air Force until the end of 1966 to learn how to employ the MiG-21 properly.[15] North Vietnam's surface-to-air missile units suffered so heavily from U.S. air attacks that for a time many units refused to fire for fear of provoking a devastating U.S. response.[16] When U.S. aircraft changed bombing targets and tactics in early 1966, the ADC concluded that "Our anti-aircraft artillery units did not alter their combat formations and their tactics in response to the new enemy tactics, resulting in many missed opportunities and a low level of combat effectiveness."[17] Another official Vietnamese history states that during this period:

> A number of anti-aircraft artillery units did not properly resolve the question of the relationship between destroying the enemy and protecting the target. They emphasized firing to scare off and drive away enemy aircraft, putting up a "curtain of fire." This type of tactic caused the rapid deterioration of gun barrels, used up large quantities of ammunition, and was not effective either in shooting down enemy aircraft or in protecting the target.[18]

Finally, during the summer of 1966, Rolling Thunder launched its first attack against a true strategic target—North Vietnam's main petroleum storage facilities—which, incredibly, had been left untouched for well over a year. The delay in attacking these facilities robbed the attack of the strategic effect it may have had earlier by allowing the North Vietnamese time to disperse their fuel stocks. A Vietnamese account says that the 11,800 tons of fuel destroyed in a 29 June attack on the Duc Giang facility near Hanoi represented only one-sixth of the facility's total fuel holdings.[19] However, the North Vietnamese acknowledged that the American attack achieved both strategic and tactical surprise.[20] The success of the attack shocked the North Vietnamese into a reassessment of the deployment and the performance of their air defense forces. One official Vietnamese account states that "The flames from the fires at the Duc Giang petroleum tank farm and our own poor performance in this battle caused much thought and severe self-criticism among commanders at all levels."[21]

The Johnson administration's reluctance to approve new Rolling Thunder bombing targets allowed the Vietnamese several months to re-deploy their forces and work out improved tactics before additional targets in the strategic Hanoi area were struck in December 1966. The strengthened Vietnamese defenses shot down thirteen U.S. aircraft in the Hanoi area in December before bad weather and new Johnson administration peace initiatives brought the attacks to a halt once again.[22] The North Vietnamese defenders were again given time to lick their wounds and correct mistakes before the attacks resumed in the spring.

By the time the penultimate Rolling Thunder battles of 1967 began, North Vietnam had perfected its integrated air defense system. This system was made up of three ADC components—Air Force fighters, SA-2 missiles, and anti-aircraft artillery guns—and a nationwide three-tiered air defense network that consisted of national-level ADC units, local air defense forces at the military region, province, and district level, and part-time militia troops who left their civilian jobs to run to their gun positions whenever the air raid alarm sounded. This integrated system, with all components operating under ADC guidance and direction, provided redundancy that allowed one element to take up the slack when another element encountered problems. When U.S. Air Force (USAF) fighters decimated the MiG-21-equipped 921st Fighter Regiment in January 1967, the ADC ordered the MiG-21s to stand down and regroup while the MiG-17-equipped 923rd Fighter Regiment stepped up operations to fill the gap.[23] In the spring of 1967, when new U.S. electronic jamming equipment rendered North Vietnam's SA-2 missiles virtually impotent against USAF aircraft, the Air Force's MiG-21 and MiG-17 force, now augmented by a second regiment of MiG-17s piloted by North Korean "volunteer" pilots, markedly increased the scale of its attacks on U.S. aircraft.[24] When anti-aircraft artillery units suffered heavy casualties in U.S. cluster-bomb attacks on their gun positions, local militia troops were used to evacuate the wounded and replace casualties on the guns.[25] By late May 1967, the Vietnamese Air Force's effort to compensate for the poor performance of SA-2 missile units had resulted in catastrophic aircraft and pilot losses. Between March and June 1967, the Vietnamese lost half their combat pilots, leaving them barely enough pilots to man a single fighter regiment. At that very moment, however, President Johnson again placed the Hanoi and Haiphong areas "off-limits" to bombing attacks, giving the entire Vietnamese air defense force time to regroup and work out new tactical and technical solutions for their problems.[26]

The North Vietnamese made maximum use of the time the United States gave them. Using new "hit and run" tactics worked out during a two-month operational stand-down, in the ten months from August 1967 to June 1968, North Vietnamese fighters shot down more U.S. aircraft (twenty-six) than they had during the previous two years of Rolling Thunder operations combined (twenty-four), and the MiG-21's kill ratio against its most capable opponent, the F-4, went from 13 to 0 in favor of the F-4 in the first half of 1967 to 5 to 1 in favor of the MiG-21 during the last half of the year.[27] North Vietnamese missile specialists and their Soviet advisors slowly and methodically developed a new missile guidance procedure to overcome the new U.S. electronic jamming equipment.[28] The new procedure, which was finally fully implemented in the late fall, was a success. During the three-month period from April to June 1967, only four USAF jets were lost to SA-2s, but in the single month of November 1967, nine F-4s and F-105s were downed by North Vietnamese missiles.[29]

Although new electronic jamming equipment introduced in December 1967 again rendered the SA-2s impotent, time was running out for Rolling Thunder. The onset of the rainy season over the Red River Delta meant that large-scale bombing of the Hanoi area had to be suspended until spring. The Tet Offensive in South Vietnam on 31 January 1968 shocked the Johnson administration to its core. On 31 March, President Johnson suspended U.S. bombing of all but the southernmost part of North Vietnam in order to begin peace talks with the North Vietnamese.

Rolling Thunder continued as an abbreviated bombing program aimed solely at Vietnamese supply lines in the narrow panhandle of North Vietnam south of the 20th Parallel. Intriguingly, North Vietnamese postwar records reveal that this restricted bombing program, which focused all the air strikes formerly spread throughout the country into this one small area, caused them more problems than had the previous nationwide bombing campaign. The heavy bombing blocked supply routes through this narrow "funnel" leading to the Ho Chi Minh Trail, and North Vietnamese supply shipments to South Vietnam through this corridor dropped from 6,500 tons in April 1968 to 1,600 tons in May, and to 1,430 tons in June.[30] By the end of 1968, the supply situation had become so bad that the flow of supplies from North Vietnam into the South had virtually stopped, and the Vietnamese were forced to reverse the direction of the Ho Chi Minh Trail to ship food and ammunition from their stockpiles in Cambodia northward back up the Trail to supply their starving troops in the northern half of South Vietnam.[31] Air defense units stationed in this heavily bombed area suffered high personnel and equipment losses that resulted in serious morale problems and widespread desertion. During the period June–October 1968, one anti-aircraft artillery regiment lost 122 men killed, 259 wounded, and 361 deserters (among the dead were five of the regiment's six gun battery commanders).[32] By this time, however, the Johnson administration had lost its stomach for the bombing, and the war, and on 1

November 1968 all bombing of North Vietnam was permanently suspended. After more than three and a half years, Rolling Thunder ended, not with a bang, but a whimper.

When the North Vietnamese launched a massive conventional ground offensive against South Vietnam at the end of March 1972, all but a tiny remnant of the massive 500,000-man American army that was fighting in Vietnam when Rolling Thunder ended had been withdrawn from Vietnam. President Richard Nixon and his administration reacted to the North Vietnamese assault on 10 April 1972 by launching a new air offensive against North Vietnam codenamed Linebacker. The North Vietnamese immediately recognized that Linebacker would not repeat the mistakes of Rolling Thunder. During the first three weeks of Linebacker, the United States blockaded all North Vietnamese ports and harbors, hit targets throughout North Vietnam with massive air strikes that included the first use of B-52s against the vital port city of Haiphong, and, in the first significant use of "smart" weapons in history, attacked key North Vietnamese supply and communications targets using laser-guided bombs. The following quote from the North Vietnamese Air Force's official history of the war demonstrates that the Vietnamese understood the improvements the United States had made in Linebacker:

> Nixon's direction of the air war was totally different that Johnson's "escalation" style of the 1965–1968 period. From the very beginning of the campaign, Richard Nixon mobilized large forces to conduct concentrated, massive attacks against the most important targets. The weapons and equipment the Americans used in this bombing campaign had been improved and were more sophisticated and much more effective than those used by the Americans in the past.[33]

The massive attacks threw North Vietnam's air defenses into complete disarray. North Vietnamese missile and anti-aircraft units were at first completely blinded by new U.S. electronic warfare equipment. During Linebacker's first B-52 strike into North Vietnam, an attack on key North Vietnamese targets in and around the city of Vinh, not only did the North Vietnamese defenders fail to engage the B-52s, their radars were so blinded by American electronic jamming that the ADC was not even able to confirm that B-52s had participated in the attack until the following day, when bomb damage assessment teams studied the bomb craters.[34] Vietnamese histories paint a bleak picture of their initial efforts against "Linebacker":

> Our air defense divisions experienced great difficulties in dealing with the enemy air force's new schemes. From 16 April to the end of May, 361st Air Defense Division [the unit defending Hanoi] failed to shoot down a single confirmed enemy aircraft. 375th Air Defense Division [assigned to protect supply lines north of Hanoi] experienced continuing problems in engaging enemy aircraft dropping laser-guided bombs. 363rd Air Defense Division [the unit defending the main port city of Haiphong] did not know how to deal with the enemy navy's false-response electronic jammers and with the continuous attacks by Shrike missiles against our radars. It was not until 24 May that 81st Missile Battalion shot down an F-8 making a night attack. This was the first aircraft confirmed to have been shot down over Haiphong during the second war of destruction [Linebacker].[35]

As for the efforts to combat the new laser-guided weapons, the Vietnamese virtually admitted defeat:

> Our air defense forces achieved poor results in the battle against laser-guided bombs. On 11 May two spans of the Long Bien Bridge in Hanoi were knocked out. On 12 May the Hoa River Bridge was destroyed. On 13 May, the "Dragon's Jaw" Bridge in Thanh Hoa province was hit by enemy bombs. By the end of May enemy laser guided bombs had destroyed 68 bridges throughout North Vietnam.[36]

The North Vietnamese Air Force, which now possessed a total of four fighter regiments (two equipped with MiG-21s, one with MiG-19s, and one with MiG-17s) was once again called on to "provide vigorous support during this time when our missile and anti-aircraft artillery troops were experiencing difficulties."[37] Although the Air Force responded with courage, initially it did not have a great deal of success. On 16 April, U.S. B-52s and tactical aircraft attacked the Haiphong and Hanoi areas. The Vietnamese account of this battle admits:

> Although we had prepared and correctly anticipated the situation, we were still confused and did not react in a timely fashion. The thirty fighter sorties flown that day (10 by MiG-21s, 14 by MiG-17s, and 6 by MiG-19s) not only did not shoot down any enemy aircraft; three of our own aircraft were shot down.[38]

On 10 May, during one of the most intense air battles of the war, the North Vietnamese admit that they lost seven MiGs, and five of their pilots were killed.[39] In an urgent meeting held on 12 May to review the situation, Air Force Headquarters decided that its tactics were outdated and that its efforts to engage in large-scale, head-to-head battles against U.S. aircraft were doomed to failure. The Headquarters decided to revert to the Rolling Thunder tactic of hit-and-run, surprise attacks, using only small numbers of aircraft in an effort to "strike a balance between destroying the enemy, protecting the target, and preserving our own forces."[40]

Using the new "guerrilla-style" tactics and ending the use of the now hopelessly outclassed MiG-17s as aerial interceptors, the North Vietnamese Air Force soldiered on through the rest of the Linebacker campaign. The MiG-19s of the 925th Fighter Regiment continued to suffer heavy losses for very little success. Even the MiG-21s of the 921st and 927th Fighter Regiments suffered significant losses, with July being an especially bad month. One of the casualties that month was Dang Ngoc Ngu, who was killed on 8 July 1972. Ngu was one of North Vietnam's leading aces (credited with seven U.S. aircraft shot down) and was the only North Vietnamese veteran of Rolling Thunder still involved in regular daylight air-to-air combat operations. Ngu may have shot down by an F-4E flown by two USAF aces, Captain Charles DeBellevue and his weapons systems officer, Captain Steve Ritchie, who claimed two of the three MiG-21s destroyed that day (the North Vietnamese admit that all three MiG-21 pilots shot down that day were killed).[41] In spite of these severe losses, North Vietnamese Air Force fighters shot down more U.S. aircraft during 1972 than during any other year of the war, including the intense air battles of 1967.[42]

Meanwhile, North Vietnam's missile and anti-aircraft artillery forces worked hard to retrain their young and inexperienced radar, missile, and gun crews (by this time, most of the veterans of Rolling Thunder had been promoted or transferred to other duties).[43] With the help of Soviet advisors, new techniques, and equipment, including optical equipment to guide the SA-2 visually, were developed and installed to overcome the new American weapons and tactics.[44] On 27 June 1972, the 57th Missile Battalion/361st Air Defense Division in Hanoi scored the first success for the new PA-00 optical target designation system for the SA-2 missile by shooting down a USAF F-4E fighter.[45]

Linebacker air attacks against the flow of weapons, supplies, and personnel being sent south from North Vietnam to sustain the Communist offensive in South Vietnam caused tremendous difficulties for the North Vietnamese and brought the offensive to a screeching halt. The Vietnamese admit that Linebacker had a severe impact on their resupply effort:

> Almost all of the important bridges on the railroad lines and on the road network were knocked out. Ground transportation became difficult. Coastal and river transportation was blocked. The volume of supplies shipped across the Gianh River to the battlefields in South Vietnam fell to only a few thousand tons per month.[46]

No matter how painful the American air attacks were, however, Linebacker did not constitute a threat to the survival of the North Vietnamese state. There was no talk in Washington of a "regime change" in North Vietnam. All the North Vietnamese had to do was hunker down, offer a few new concessions at the negotiating table, and wait the Americans out. They knew the Americans were eager to make a swift exit from the seemingly interminable war in Vietnam. As a result of a new North Vietnamese "peace" initiative in Paris, a tentative peace agreement was reached in October 1972, and on 23 October all bombing of North Vietnam north of the 20th Parallel was suspended. Nothing was easy about the Vietnam War, however, and the same was true of the effort to end it. While Linebacker was essentially over, the war had one last spasm before it ended.

When the tentative peace agreement broke down because of South Vietnamese intransigence and North Vietnam's revocation of several of its earlier concessions, the Nixon administration decided to break the impasse with one final application of overwhelming air power. On 18 December 1972, President Nixon ordered the initiation of Linebacker II, a short, intense bombing campaign using more than 200 B-52s in massed attacks against hitherto inviolate targets in Hanoi and Haiphong. While Nixon's true purpose in launching the Linebacker II attacks continues to be a subject of debate, the most accurate description of the purpose of the campaign may have been a jocular statement by John Negroponte, Henry Kissinger's Vietnam specialist at the Paris Peace talks. Speaking of Linebacker II, Negroponte quipped, "We bombed them into accepting our concessions."[47]

The Vietnamese Air Defense Command had spent years studying how to engage and destroy B-52s. As far back as 1966, a missile regiment had been sent to the southernmost part of North Vietnam to study and test ways to combat B-52s.[48] In 1971, the Vietnamese Air Force sent MiG-21s to the area just north of the Demilitarized Zone to test tactics for intercepting B-52s. No B-52s were shot down by either the missiles or the MiGs, although the heavy bombers had several close calls. Many North Vietnamese officers began to express doubt that their country was capable of shooting down a B-52.[49] After a concerted effort throughout 1972, in November 1972 a special research team working with the 263rd Missile Regiment in southern North Vietnam finally succeeded in shooting down a B-52. The aircraft, crippled by a missile explosion, crashed in Thailand as it tried to limp back to its base.[50] This success came just in time. On 24 November, the ADC presented its plan for defending against a projected U.S. B-52 attack on Hanoi and Haiphong to the General Staff for final approval. The General Staff approved the plan and ordered the ADC to complete all preparations and be able to implement this plan by 3 December.[51] The first B-52 attacks of Linebacker II were carried out on the night of 18–19 December, only fifteen days later. Using the new techniques and tactics laid out in the plan, North Vietnamese SA-2 missile units shot down fifteen B-52s and damaged nine more during the two-week bombing campaign.[52]

Both the American and the Vietnamese sides claimed victory when Linebacker II ended on 30 December: the Americans because the North Vietnamese agreed to a number of concessions in the final Vietnam Peace Agreement signed on 27 January 1973 and because the North Vietnamese returned all American prisoners of war; and the North Vietnamese because they had shot down so many B-52s and because the Americans agreed to a permanent and unilateral withdrawal of all U.S. military personnel from Indochina as part of the peace settlement. Some people have asserted that North Vietnam could have been forced into total and complete surrender with just a few more days or weeks of bombing, but this is extremely doubtful.[53] While the North Vietnamese admit that their forces suffered heavily during the bombing, they had survived, and they had destroyed so many U.S. aircraft that they called the battle a "Dien Bien Phu in the air." The battle is still celebrated in Hanoi every year as a great victory. It is also clear that the North Vietnamese were aware of the powerful U.S. domestic political pressure on the Nixon administration, both in the Congress and among the general public, to get out of Vietnam. The North Vietnamese knew as well as did President Nixon that, in Nixon's words, "If we did not end the war by concluding an agreement at the next Paris session, then when Congress returned in January it would end the war by cutting off the appropriations."[54]

The bombing and the war ended in the same way they began: in confusion and controversy, with the two sides not even able to agree on who had won and who had lost.

Notes

A shorter version of this article was published in the February 2005 issue of *Vietnam* magazine under the title "American Air Power: The View from North Vietnam."

1. Chris Hobson, *Vietnam Air Losses: United States Air Force, Navy, and Marine Corps Fixed-Wing Aircraft Losses in Southeast Asia 1961–1973* (Hinckley, UK: Midland Publishing, 2001), 268–70.

2. Le Duan, *Thu Vao Nam* (Letters to the South) (Hanoi: Su That Publishing House, 1985), 50–66; Military History Institute of Vietnam, *Victory in Vietnam: The Official History of the People's Army of Vietnam, 1954–1975*, tran. Merle L. Pribbenow (Lawrence: University Press of Kansas, 2002), 74–75, 124–25.

3. Senior Colonel Va Van Minh (Chief Editor), *Quan Khu 4; Lich Su Khang Chien Chong My Cuu Nuoc (1954–1975)* (Military Region 4: History of the National Salvation Resistance War Against the Americans, 1954–1975) (Hanoi: People's Army Publishing House, 1994), 105–10; Bui Tin, *From Enemy to Friend* (Annapolis, Md.: Naval Institute Press, 2002), 38–42; Qiang Zhai, *China and the Vietnam Wars, 1950–1975* (Chapel Hill: University of North Carolina Press, 2000), 130–33.

4. Air Defense Service, *Lich Su Quan Chung Phong Khong, Tap I* (History of the Air Defense Service, Volume I) (Hanoi: People's Army Publishing House, 1991), 195.

5. Ibid., 196–97.

6. Major General Nguyen Xuan Mau and The Ky, *Bao Ve Bau Troi: Hoi Ky* (Guarding the skies: A memoir) (Hanoi: People's Army Publishing House, 1982), 53–59.

7. General Oleg Sarin and Colonel Lev Dvoretsky, *Alien Wars: The Soviet Union's Aggressions against the World, 1919–1989* (Novato, Calif.: Presidio Press, 1996), 91; Segei Blagov, "Missile Ambushes: Soviet Air Defense Aid," *Vietnam* magazine (August 2001), 28.

8. Military History Institute, 165.

9. Ibid., 164–65; Air Defense Service, *Lich Su Quan Chung Phong Khong, Tap II* (History of the Air Defense Service, Volume II) (Hanoi: People's Army Publishing House, 1993), 69.

10. Sarin and Dvoretsky, 91–95; Blagov, 27; Zhai, 133–35; Senior Colonel Nguyen Van Minh, ed., *Lich Su Khang Chien Chong My Cuu Nuoc, 1954–1975, Tap V* (History of the Resistance War Against the Americans to Save the Nation, 1954–1975, Volume V) (Hanoi: National Political Publishing House, 2001), 270–71.

11. Air Defense Service, 2:8.

12. Le Nguyen Ba (Chief Editor), *Lich Su Su Doan Phong Khong Ha Noi* (History of the Hanoi Air Defense Division) (Hanoi: 361st Air Defense Division, 1985), 37–38.

13. Ibid., 38–39.

14. Colonel Ta Hong, Lt. Col. Vu Ngoc, and Lt. Col. Nguyen Quoc Dung, *Lich Su Khong Quang Nhan Dan Viet Nam (1955–1977)* (History of the People's Air Force of Vietnam, 1955–1977) (Hanoi: People's Army Publishing House, 1993), 121.

15. Nguyen Trong Thuan (Chief Editor), *Lich Su Su Doan Khong Quang 371* (History of the 371st Air Force Division) (Hanoi: People's Army Publishing House, 1997), 54–70.

16. Nguyen Xuan Mau and The Ky, 133.

17. Air Defense Service, 2:80.

18. Military History Institute, 187.

19. Cao Hung, ed., *Thu Do Hanoi: Lich Su Khang Chien Chong My Cuu Nuoc, 1954–1975* (The Capital, Hanoi: History of the National Salvation Resistance War Against the Americans, 1954–1975) (Hanoi: People's Army Publishing House, 1991), 112–13.

20. Air Defense Service, 2:88.

21. Le Nguyen Ba, 56.

22. Hobson, 82–83.

23. Ta Hong, 155–56. A Vietnamese fighter regiment possessed a total of 30–36 aircraft.

24. Marshall L. Michel, *Clashes: Air Combat Over North Vietnam, 1965–1972* (Annapolis, Md.: Naval Institute Press, 1997), 91; Air Defense Service, 2:123; Nguyen Van Minh, 271. According to BBC and Reuters wire service reports on 31 March 2000, fourteen North Korean personnel, including eleven pilots, were killed in action in North Vietnam during the war.

25. Air Defense Service, 2:131, 147, 168.

26. Ta Hong, 168–69; Wayne Thompson, *To Hanoi and Back: The U.S. Air Force and North Vietnam, 1966–1973* (Washington D.C.: Smithsonian Institution Press, 2000), 69, 74–75.

27. Hobson, 270; Michel, *Clashes,* 139–40.

28. Merle Pribbenow, "The –Ology War: Technology and Ideology in the Vietnamese Defense of Hanoi, 1967," *Journal of Military History* 67 (January 2003):189–93.

29. Hobson, 270.

30. Air Defense Service, 2:257.

31. Truong Son (penname of Senior Colonel Pham Te), *Nhung Nam Thang Soi Dong Nhat Tren Duong Ho Chi Minh* (The hottest period on the Ho Chi Minh Trail) (Ho Chi Minh City: Ho Chi Minh City Publishing House, 1994), 13–14.

32. Air Defense Service, 2:268–69.

33. Ta Hong, 234.

34. Air Defense Service, *Lich Su Quan Chung Phong Khong, Tap III* (History of the Air Defense Service, Vol. III) (Hanoi: People's Army Publishing House, 1994), 93–94.

35. Ibid., 118. The Air Defense Command had two categories for recording aircraft shot down: "shot down" and "confirmed shot down" (literally: "shot down on the spot"). The "shot down" category was tremendously inflated as it recorded virtually every claim made. This category seems to have been used primarily for propaganda and morale-building purposes. The requirements for recording an aircraft as "confirmed shot down" were more stringent and required outside confirmation (wreckage, reports from other units/observers, U.S. public acknowledgement of losses, and so on. While there were many mistakes and some exaggerations in this category as well, the "confirmed shoot down" records were much more accurate when compared with U.S. loss records. See Marshall Michel, *The Eleven Days of Christmas* (San Francisco: Encounter Books, 2002), 296.

36. Michel, *Eleven Days,* 119.

37. Ibid., 116.

38. Nguyen Trong Thuan, 152–53.

39. Ibid., 161.

40. Ta Hong, 258; Nguyen Trong Thuan, 162–63.

41. Nguyen Trong Thuan, 164–69; Michel, *Clashes,* 244–45.

42. According to Hobson, 271, thirty U.S. aircraft were shot down by MiGs in 1972. Michel, *Clashes,* 277, gives a figure of twenty-eight. North Vietnamese Air Force claims of aircraft shot down are much higher and are not reliable.

43. Air Defense Service, 3:108.

44. Ibid., 119.

45. Le Nguyen Ba, 128; Hobson, 230.

46. Military History Institute, 300.

47. Society for the Historians of American Foreign Relations, March 2002 Newsletter, accessed at http://shafr.history.ohio-state.edu/Newsletter/2002/MAR/LETTERS.HTM on 21 Apr. 2003.

48. Air Defense Service, 2:243.

49. Military History Institute, 315.

50. Ibid., 317; Hobson, 240.

51. Air Defense Service, 3:179.

52. Michel, *Eleven Days of Christmas,* provides an excellent and detailed account of the Linebacker II campaign from both the Vietnamese and the American sides.

53. Air Defense Service, 3:179.

54. Richard M. Nixon, *RN: The Memoirs of Richard Nixon,* vol. 2 (New York: Warner Books, 1979), 230.

Jet Aircraft and Defense Imperatives, World War II to Vietnam

Thomas A. Keaney

At the end of World War II, the commanding general of the U.S. Army Air Forces, General Henry "Hap" Arnold, asked Dr. Theodore von Karman, a prominent scientist in the field of aeronautics, to lead a special study. Arnold wanted von Karman to organize a study of technologies of potential use to the Army Air Force in the future. Von Karman, working with the group of thirty-six scientists, examined the lessons of World Wars I and II, traveled to Europe to consult with scientists in Great Britain, France, and the by-then defeated Germany, and examined the products of their laboratories. Von Karman and his group's work, under the title, *Toward New Horizons,* would eventually lead to thirty-two separate monographs written by twenty-five authors. Von Karman's introductory volume, *Where We Stand,* set forth the summary in which he listed the "fundamental realities" of future air warfare. He concluded that aircraft would fly beyond the speed of sound, face defenses by target-seeking missiles, require extreme speed to penetrate enemy defenses, and have perfect communications between aircraft and commanders. He also noted that unmanned devices would travel thousands of miles to targets, with small amounts of explosive that could destroy several square miles of area. And, with precise navigation and communication independent of weather, airborne task forces could strike distant points and be supplied by air. He particularly emphasized jet, rocket, and atomic propulsion for aircraft and urged study of two other fields of research, tailless aircraft and radar.[1]

Looking back on von Karman's projections after some sixty years, two aspects in particular stand out. First, the report captured quite accurately the technical factors that were to govern air power's development in the years ahead. Second, no matter how well known the technical aspects were perceived, developing operational systems, even in a still-imperfect way, has taken much effort, false starts, and unplanned diversions. Knowing the trends and possibilities in technology set out the direction, but many operational, budgetary, and political imperatives along the way determined the specific courses taken, and those courses not taken.

World War II had brought rapid advances in combat aircraft capabilities, and the progress continued after the war's end in 1945. In fact, the exploitation of many technologies that had seen their birth in the war proceeded at perhaps an even faster pace, despite drastic reductions in the budgets. And the results were dramatic. By 1950, even the most advanced of the World War II aircraft had become obsolete. By 1955, a further generation of aircraft had themselves far surpassed those higher performance limits. Still, those experiencing the rapid changes had seen nothing like

163

what was soon to come: space flight, intercontinental ballistic missiles (ICBMs), airborne guided missiles, and electronic warfare. These all arrived, not a generation later, but within the *next* five years. Within the careers of many then-serving officers, many of the conditions cited had been realized. Jet aircraft helped usher in an explosion of technological change that brought as many as six generations of fighter aircraft in thirty years.[2] The terms "technological revolution" or "transformation," if accurate to describe current conditions, definitely apply to those post–World War II years when a confluence of factors, technological as well as political, ushered in a new age of combat aircraft and much else.

This paper examines several of these fundamental realities, particularly jet aircraft development, and the technologies that brought rapid development of combat aircraft. The military competition that took place in those years involved much more than a sequenced expansion of new and developing technologies for aircraft. It involved a series of factors, strategic, tactical, as well as technological, and the paths had to deal with many uncertainties, some mis-steps, and some very unexpected, or unforecast, developments. That after all has been the history of aviation for one hundred years, and we are likely to see a continuation of such events.

Because the subject is so large, let me bound this account in time and focus. This paper looks at roughly the twenty years from the end of World War II, through the initial stages of the Cold War, through the Korean War, to the 1960s, when the United States began to learn a further set of air power lessons in the Vietnam War. The focus is technological change, with the development of the turbojet engine being a centerpiece of the technologies, but technology did not act alone. The Soviet competition, limited defense budgets, and scarce materials also played significant parts in this history. This account examines the factors of change as they involved most directly fighters and bombers, recognizing that the stories of helicopters, airlift, electronic warfare aircraft, among other categories of aircraft, both played a part and saw radical shifts themselves. The terms "fighter" and "bomber" need further definition, and I will deal with those definitions in context during the analysis. And finally, this account looks only at the developments in U.S. manned combat aircraft, a vital component, but far from the entirety of aviation's military revolution in this period. For instance, one could with equal justification cite the rapid developments in ballistic missiles, in space exploitation, or in other unmanned space and airborne systems as major elements in this revolution.

World War II itself contained many examples of the rapid pace of aircraft development that characterized the periods before and afterwards. Front-line aircraft employed in the skies over Europe and the Pacific (namely, American fighters—P-47 Thunderbolts, P-51 Mustangs, P-38 Lightnings, F-4U Corsairs, F-4F Wildcats, and F-6F Hellcats, the most well-known, and heavy bombers—the B-17 Flying Fortresses, B-24 Liberators, and B-29 Super Fortresses) represented designs that sought the maximum advantages within the critical factors of speed, range, and armaments (or bomb load). Tradeoffs took place between these critical factors, and experience showed that even a slight edge in performance brought great tactical advantages. Advantages over enemy aircraft proved to be fleeting, however. Fighters of 1939 had great difficulty competing with the aircraft introduced only two or three years later. And, operational requirements could dictate additional factors. The B-29, for instance, not only flew 70 mph faster than the B-17 and carried a far larger payload, its greater range, double that of the B-17, made possible a strategic bombing campaign against Japan, a mission impossible for the B-17, given the great distances involved in Pacific theater operations.

As new and remarkable as the B-29 performance might have appeared, it represented only a waypoint. Under development in the United States during the war was an even bigger and longer range aircraft, the B-36, faster and with nearly double the range of the B-29. The B-36 would permit a bombing campaign against either Germany or Japan while flying from bases in the United States. The rate of progress meant that planners had to cope with the possible obsolescence of new aircraft designs after only five years. In other words, when the first production models of an aircraft began appearing, designers had to have had already planned the next generation.

One might have assumed that with the war's end the competition engendered by the war and the remarkable process of change would at least attenuate. Other factors, technological and strategic, however, took over to actually accelerate the process after the war. First, the appearance of the ME-262, a German twin-engine jet fighter, in the skies over Europe in the fall of 1944 signaled the new technological competition and in the process threw a fright into the British and American military leadership. The great speed advantage of these aircraft over any other fighter made them a significant threat to the bomber formations and, at the time, there were fears that these aircraft by themselves might threaten the actual continuation of the bombing campaign. The commander of U.S. strategic air forces, General Carl Spaatz, advised that the Germans could have hundreds of jet aircraft operational before any American jets would be ready, made diversions in the bombing campaign specifically to target German jet assembly plants and associated facilities as one of the highest priorities.[3] In the end, the small numbers of the ME-262s available limited their impact, but the point was made in the military planners' mind of how easily a sudden technological advantage might reverse an air campaign's outcome.

The German jet-engine program had led the way, but British and American planners had recognized the jet's potential and had begun work on engine and aircraft designs early in the war. The British, continuing their prewar research, produced the Gloster Meteor by the summer of 1944 and put it to work soon after over Great Britain flying air defense missions against V-1 rocket attacks.[4] The first American production jet aircraft, the Locked P-80, began design in 1943 and first flew in January 1944. Even then, it owed its engine design to the British and their research. The P-80 did not arrive in operational service until the fall of 1945, too late for the war, but not for lack of trying: by February 1945, the P-80 had gained top priority for production along with the B-29.[5] By then, and before the war's end, the United States had begun moving rapidly, ordering development tests on eight different types of jet propulsion and contracting for four other models of jet fighters and three jet bombers. Most of these designs fell short of expectations and did not enter the inventory, but two were to gain particular prominence later, the B-47 and the F-86.

Even in the infancy of jet turbine design, jet-engine aircraft showed they could outpace even the fastest piston aircraft, and further progress in designs promised still more in performance. With all the advantages, one aspect of jet-engine performance served to limit their application—fuel consumption. Not only could jet aircraft fly fast and high, they had to for the sake of fuel economy. In the early stages of jet design, planners projected a jet aircraft would have to attain 600 mph to be as economical on fuel as a piston aircraft. And, while jet engines operated more efficiently at high altitude—the reverse condition of piston engines—they had questionable usefulness for aircraft whose roles required low-altitude operations.

In the postwar period, however, the emphasis was on speed, not low-altitude flight, and here the jet excelled. Most prominently, jet-engine propulsion seemed capable of removing the then-assumed absolute cap on aircraft speed, the speed of sound—760 mph at sea level—known as the sound barrier. In 1945, much was unknown about the aerodynamics of flight faster than the speed of sound. During flight tests and aircraft in high-speed dives during air combat, pilots could lose control of the aircraft as shock waves built up on control surfaces, effectively locking the controls. Tests showed that as aircraft approached that speed, drag rose, propeller efficiency declined, and flight controls became unreliable or experienced a series of unexpected characteristics, with a consequent loss of control, and often loss of the aircraft.[6] German research had studied these phenomena, and at the war's end, American designers gained access to German aircraft, designers, and testing results, including wind-tunnel data. They learned that the Germans had made significant advances beyond the United States and British designers in a number of areas. In several key areas, the German research provided was particularly beneficial: in swept-wing planforms for high-speed aircraft, cruise and ballistic missiles, and surface-to-air missiles, and in understanding the phenomena of transonic and supersonic flight. The lessons learned brought immediate changes in U.S. designs then in progress, most notably changes in the Boeing XB-47 and North American XP-86 to swept-wing designs.[7]

In addition to the technological changes underway, two further developments had a significant impact: the transformation of the Soviet Union from wartime ally into postwar enemy, and the development of the atomic bomb and its use against Japan. Both events had profound implications for the types and planned employment of aircraft developed, as well as the pace of that development. The shift of the Soviet Union from troublesome ally, to hostile competitor, to bitter enemy, took place within five years of the end of the war. The details of this shift from alliance to Cold War are well known and have no need of repeating here. The impact of this shift meant that a new technologically advanced hostile country now competed in all military areas with the United States, particularly in air and later in space power. In World War II, the Soviets had no aircraft that could match the British and American designs, but the Soviets gained tremendous benefit for their research and development programs in

1945 by getting control of much of the German aircraft and research data. How quickly they adapted these designs for their own use became evident in the Korean War.

The atomic bomb itself had dramatic effects on air power developments, effects further magnified when both the Soviets and Americans had attained stockpiles of nuclear weapons. The implications of the bomb reopened the debate on the value of strategic bombing. Until the atomic bomb's use in 1945, the value and effects of strategic bombing remained controversial. Bombers had not lived up to many of the prewar expectations in a number of ways, including bomber vulnerability, problems with bombing accuracy, and selection of targets. Now, at one stroke, many of the problems had seemingly been swept aside. Atomic bombs promised the ability to inflict devastating destruction through the flight by a single bomber, and with no need for increased accuracy. That premise was to lead to a prime emphasis on development of the heavy, long-range bombers to deliver these weapons. Later in the 1950s, when the size and weight of nuclear weapons permitted, the same primacy of nuclear weapons delivery also guided the design of fighter aircraft. Similarly, the Soviet production of long-range bombers to target the United States with nuclear weapons caused a resulting emphasis of U.S. long-range fighter interceptors.

As a result of the rapidly changing landscape of international affairs, the immediate postwar years found the United States both uncertain of its policy toward the Soviet Union and of the use of the atomic bomb in its military plans. In addition, military reorganization preoccupied planners of the time, plans that led to the establishment of the U.S. Air Force in 1947 and a necessary sorting out of roles in aviation among the services. Strategic Air Command (SAC) had already been established a year earlier in 1946 and took a major role in determining long-range bomber development. That time period coincided with the growing recognition of the aggressive intentions of the Soviet Union and the consequent drive by the United States to meet this growing and more evident threat. Long-range bomber development assumed center stage in the buildup to meet this threat.

Bombers

In 1947, George Kennan, a U.S. diplomat just returned from Moscow, wrote his famous "Mr. X" article laying out the basis of Soviet expansionism and of the need for a U.S. policy of containment. That same year, the United States came to the aid of Greece and Turkey, then being endangered by Soviet-sponsored guerillas, and announced the Marshall Plan to assist in the economic recovery of Europe. At that time, the United States had two major bomber programs under development, both begun during World War II: the B-36 with six piston-engines (with pusher propellers mounted on the trailing edges of the wings), whose design dated to 1941, and the B-47, a six jet-engine bomber. These aircraft differed considerably, but both proved necessary in the international political climate of the times. The B-36 served as a back-up plan in case Great Britain fell to a Soviet attack and the United States needed the aircraft's great range to attack Germany.[8] By 1947, however, though it remained in development, the B-36's lack of speed made it vulnerable to air defenses of jet fighters. Without a suitable substitute, however, in the short term it served as a stop-gap long-range bomber. Its projected combat range of 8,000 miles gave it essentially twice the range of the B-29 or the planned-for B-47.

As early as 1944, the Army Air Force issued design requests for the B-52 as a successor to the B-36; the request came, remarkably enough, over two years before the B-36 had even flown its first test flight.[9] Thus, though planned to be replaced as quickly as possible, the B-36 was to become the primary U.S. long-range bomber (the B-29 and its mate the B-50 were redesignated medium bombers). Though at times close to cancellation, the B-36 program endured principally because of the growing Soviet threat. In 1948, for instance, when cost overruns and difficulties with its engines' design threatened the program, the Soviet Union's blockade of Berlin accentuated the need for producing bombers as soon as possible. Modifications, whenever possible, took place. To increase its speed, the B-36 had four jet engines added, J-47s, the same ones then on the B-47, increasing its maximum speed from 381 to 439 mph.[10]

The aircraft faced much debate on its performance throughout its development and service. The most notable public dissent came as part of what was termed the "Revolt of the Admirals" in 1948, when as part of a reaction to the cancellation of the planned aircraft carrier *United States,* the Navy contested the capabilities of the bomber. In highly publicized hearings before the House Armed Services Committee, the Air Force had to defend itself against Navy charges that the B-36 was obsolete, particularly in light of the appearance of the Soviet MiG-15 jet fighter at that time. The Air Force's defense of the bomber was a difficult one, given that the Air Force had essentially come to the same conclusion and was simply in the process of buying time until a successor was developed. Despite its shortcomings, the aircraft went into production, with the first operational units appearing in 1950 and stayed in production until 1954, the time that B-52s were planned to begin production. Three hundred and sixty-six B-36s were produced, both bomber and reconnaissance versions, with the last being retired in 1959. Though there were many plans for its continued use, including converting to turboprop engines and making it a cruise missile carrier, the B-36 had come to be seen as an interim measure, not suitable in an age when speed was the all-important concern.[11]

The B-47, the B-36's near contemporary, could not have differed more from it. Where the B-36 was a clear continuation of World War II bombers—it resembled an elongated B-29—the B-47 took the very different design of a jet aircraft. It had swept wings, a sleek, slender fuselage, representing a big step in aerodynamic design over its predecessors. Its major drawback, one it shared with all early jet aircraft, was its range, only 2,000 miles. Though the aircraft had met the design range identified for it, the distances to Soviet targets had made this earlier measure now insufficient.

Despite its range limitations, the B-47's speed made it a highly valued aircraft, and the Air Force sought to rush it into production as rapidly as possible. Again, the Soviet MiG-15 became the key factor in the accelerated development. The Soviet Union had moved ahead rapidly in jet engine and aircraft design and unveiled the new jet interceptor in December 1947. The swept-wing MiG could fly at 675 mph and attain an altitude of 50,000 feet.[12] (The United States had an aircraft to match this capability, the F-86, to be discussed subsequently). The MiG-15 created two problems for U.S. aircraft operating against it. First, the MiG put at risk the B-36's capability to operate in Soviet airspace, though the B-36 had the range for the task. The B-47's speed could neutralize the MiG's capabilities, but its range was a severe drawback in reaching Soviet targets.

The Air Force proposed several alternatives to cope with the MiG threat and the differing capabilities of their bombers. One concept entailed carrying a small fighter aircraft inside the B-36. A small aircraft designed by McDonnell Aircraft, the XF-85 Goblin, was to be launched to defend the bomber against airborne threats and then be retrieved subsequently by the B-36.[13] That program produced two such parasite aircraft, but ended before any airborne tests of the concept took place. An alternative design to be employed by the reconnaissance version of the B-36 was to carry an RF-84F fighter under the RB-36 fuselage, launch the fighter to penetrate enemy airspace and then return to the mother ship. The early version of a standoff weapon delivery technique made it as far as forming a squadron of modified F-84Fs and a dozen B-36s modified as carriers.[14]

Attempts to extend the range of the B-47 through redesign in most cases involved compromising its speed, a solution with its own disadvantages. SAC planners therefore looked to other measures. One proposal, showing perhaps their desperation in attaining greater range, involved a maneuver that would have B-47s as well as other bombers link up, joined wing tip-to-wing tip after takeoff, and fly so-joined until approaching enemy territory; the aircraft would then separate for their independent bombing missions. The concept for this procedure originated from the principle that aircraft in linked formation would have less induced drag, and thus more lift and efficiency of flight. The data cited was for an increase in range of over 50 percent for six to eight aircraft joined in that way. The plan involved modifying aircraft with so-called "floating wing tips" to facilitate this procedure. First proposed in February 1949, the concept was still under review over a year later, to the extent that SAC recommended to the Air Force's Senior Officers Board in April 1950 that these floating wing tips be developed as soon as possible.[15] Whether any in-flight tests of this procedure took place is not recorded, but the execution of such a procedure would appear to require a remarkable degree of coordination among the crew members, not to mention even greater synchronization of the autopilots and flight controls on the aircraft involved. The same principle of floating wing tips, in a modified form, became one measure proposed in the design of the XB-70.

If the plans for creating an effective long-range bomber force in 1950 involved a measure of desperation, there were ample reasons for it. Two events in 1950 help to provide this context. In the spring of 1950, National Security Council (NSC) paper-68 appeared. This special study, directed by President Harry S. Truman and produced by the Defense and State Departments, gave an assessment of the Soviet threat and of U.S. actions to counteract it. NSC-68 described an ominous threat of Soviet military power and recommended massive increases in U.S. defense spending. Outlining Soviet military capabilities, the report stated that should a war occur in 1950:

> The Soviet Union and its satellites are considered by the Joint Chiefs of Staff to be in a sufficiently advanced state of preparation immediately to undertake and carry out the following campaigns.
>
> a. To overrun Western Europe, with the possible exception of the Iberian and Scandinavian Peninsula; to drive toward the oil-bearing areas of the Near and Middle East; and to consolidate Communist gains in the Far East;
>
> b. To launch air attacks against the British Isles and air and sea attacks against the lines of communications of the Western Powers in the Atlantic and the Pacific;
>
> c. To attack selected targets with atomic weapons, now including the likelihood of such attacks against targets in Alaska, Canada, and the United States. Alternatively, this capability, coupled with other actions open to the Soviet Union, might deny the United Kingdom as an effective base of operations for allied forces. It should also be possible for the Soviet Union to prevent any allied "Normandy" type amphibious operations intended to force a reentry into the continent of Europe.
>
> . . . During the course of the offensive operations listed in the second and third paragraphs above, the Soviet Union will have the air defense capability with respect to the vital areas of its own and its satellites' territories which can oppose but cannot prevent allied air operations against these areas. . . .
>
> . . . It is believed that the Soviets cannot deliver their bombs on target with a degree of accuracy comparable to ours, but a planning estimate might well place it at a 40–60 percent of bombs sortied. For planning purposes, therefore, the date the Soviets possess an atomic stockpile of 200 bombs would be a critical date for the United States, for the delivery of 100 atomic bombs on targets in the United States would seriously damage this country.

The report went on to state that a Soviet stockpile of 200 atomic weapons would be in place by mid-1954.[16]

The implications of an attack such as posited by NSC-68 had direct implications for the U.S. military and for bomber operations: no advanced bases could be relied on, either in Europe, the Middle East, or in Asia, thus severely curtailing the effectiveness of medium-range bombers such as the B-47; air attacks into the Soviet Union or its satellites would be met by capable air defenses, thus the danger to the B-36; and, the United States would have to be prepared to counter these conditions as early as 1954, emphasizing the need for rapid rearmament.

Two months after the publication of NSC-68, the North Korean army invaded the South, initiating the Korean War. Seen through the lenses of NSC-68, this attack was perceived as an opening move by the Soviet Union to engage in a worldwide war with the United States and its Western allies in Europe. Those who would have contested the projections of NSC-68 and its budget implications now had second thoughts. The defense budget increased fourfold between 1950 and 1954, and getting bombers in production and into operation as soon as possible assumed an even higher priority.

While the Korean War spiked development of bombers for nuclear weapons roles against the Soviet Union, the new bomber fleet played no role in it. Seen as war

of limited nature and with few strategic targets for bombers, only B-29s not configured for carrying nuclear weapons served in the Korean theater; B-36s and B-47s had
become part of the nuclear strike force aimed at the Soviet threat and played no role
in Korea whatsoever.

Meanwhile, actions continued to increase the range of the nuclear strike force.
Those familiar with current air operations know the solution found for extending
aircraft range: in-flight refueling by tanker aircraft specially configured for the role, a
common enough procedure now, but a major undertaking at the time. Though experiments with aerial refueling had taken place in the United States dating back to the
flight of the *Question Mark* in 1929,[17] developing the systems and techniques needed
for conducting aerial refueling operations of jet aircraft on a sustained basis represented an enormous endeavor, requiring a significant investment in people, aircraft,
and training. The C-97, a propeller-driven World War II-era cargo aircraft—essentially a modified B-29—fitted with aerial refueling systems became the initial tanker
aircraft purchased, the KC-97. Initially tied to B-47 production, in 1950, SAC recommended to the Air Force Senior Officers Board that one KC-97 be procured for each B-
47 programmed.[18] Eventually, over 800 KC-97s entered the Air Force inventory.[19]
Operating aircraft with such different performance characteristics brought its own
difficulties, and by the late 1950s, the KC-97s were supplanted by KC-135s, an aircraft
still in use and a far better match for the B-47, B-52, and the bombers that followed.
Begun as a means to extend the range of bombers for strikes into the Soviet Union,
later the aerial tanker's role expanded to become a part of routine flight operations for
fighter, cargo, and many other types of aircraft in the Air Force, Navy, and Marine
Corps.

The operational life of the B-47 lasted from 1952 to 1966, seven years after the last
B-36 retired, and a relatively long tenure for most jet aircraft of the period. The production of over 2,000 took place between 1952 and 1957 (all versions, including over
200 reconnaissance, RB-47s). The influence of the Korean War and the predicted
growth of the Soviet threat spurred its rapid production; the Air Force made arrangements to open production lines for the Boeing aircraft product at factories run by
Lockheed and Douglas.[20] Inevitably, perhaps, the acceleration of the bomber's production meant breakdowns in coordination, continual engineering changes, slips in
schedule, and additional expense.[21] During much of the 1950s, however, until the
arrival of the first B-52s, the B-36s and B-47s, one model the last of the old designs,
the other the first of the new, carried the load of the SAC bombing mission.

Development of the B-52 trailed that of the B-47 by only a few years. Seen as the
successor to the B-36, the B-52 design began as a straight-wing turboprop aircraft, a
design in keeping with the range sought, over 10,000 miles non-refueled. In 1948,
however, Boeing converted the design to a turbojet—employing the next generation
of engines, the J-57, borrowed aircraft design ideas from those of the B-47, and
promised the Air Force delivery of a bomber of long range (over 8,000 miles) and high
speed (over 570 mph).[22] That design emerged as the leading contender for carrying
out the long-range bomber role, and the need for such a bomber rapidly came to the

fore. In the midst of the Korean War emergency in the winter of 1950–51, during the same period in which the B-47 development was accelerated, that the Air Force issued orders for production of B-52s. One other aircraft served as serious competition with the B-52 design, the Northrop XB-49 flying wing. A redesign of an earlier propeller version, the XB-35, this eight jet-engine aircraft promised both high speed and long range, but flight tests during the crucial period of 1948 showed a series of stability and flight control problems, including a crash and killing of the crew in June 1948.[23] Unfortunately for the program, at the time the low radar signature features of the aircraft were not one of the characteristics prized in the competition, and the program was cancelled in 1949. However, the recognized value of stealth technology led to the development of its descendant, the B-2, over thirty years later.

The B-52 went into major production in 1954, with the first operational units arriving in 1956. Production continued until 1962, with a total of 744 built. Most of the latest version of the B-52, the "H" model, delivered in 1961 and 1962, are still in service over forty years later and have an estimated service life to beyond 2040.[24]

That the B-52 remained on active service a half-century after its inception would have astonished any observer of aircraft advances in the 1950s. At that time, the service life for combat aircraft was often no longer than ten years. The B-52 and B-47 had replaced aircraft (B-36 and B-29) less than ten years old at the time, and they themselves were in line for replacement on a like time schedule. A significant factor in determining the replacement schedule of these bombers was the relative capabilities of the Soviet air defense they would have to face. The United States assumed, even when the B-52 requirements were issued, that in the 1955–60 time period, those defense would include supersonic all-weather interceptors and guided surface-to-air missiles. And, starting in 1956, U-2 over flights accumulated evidence of the growing capabilities of Soviet air defenses, the best evidence coming in 1960 when a surface-to-air missile shot down a U-2. By then, SAC had determined that aircraft penetrating Soviet airspace could do so successfully only at low altitude.[25]

Identification of the high altitude threat caught planners in mid-stride, however, since the short-cycle time between bomber development meant that while the B-52s and B-47s were still in production, plans were underway already for two new bombers that would fly faster and higher. The Convair B-58, a medium-range bomber and B-47 replacement, had a delta wing, flew at Mach 2 (twice the speed of sound), and incorporated radically new avionics and flight control systems. The aircraft gave SAC a supersonic bomber, but at great cost and, as it turned out, with marginal utility. Testing took years more than planned, and finally SAC only reluctantly allowed production of the aircraft because of the money already expended. An originally planned production of nearly 300 aircraft was reduced to 86, 20 percent of which crashed, and their operational life lasted less than a decade, 1961 to 1970.[26] Meant as a replacement for the B-47, it lasted in service only four years longer than the B-47.

The XB-70, the planned replacement of the B-52, faced even greater difficulties. Programmed to have the speed of the B-58 and range of the B-52, the designs for the XB-70 promised even more than that—Mach 3 at an altitude of 75,000 feet. Such a design required the realization of leaps in technology in several areas, special manu-facturing techniques, and a radically different suite of avionics. Few of these require-ments came to fruition on time or at all. The first flight scheduled for 1960 slipped eventually to 1964, and costs ballooned to the point that it was to cost ten times that of a B-52. Though it did prove capable of Mach 3 flight at 75,000 feet, its other characteristics were much more limiting. By the 1960s, bombers could survive only by flying at low level over the Soviet Union, and the XB-70's design became a liability in those circumstances.[27] The XB-70 program ended with two-prototype aircraft, leav-ing the long-range bomber mission to the B-52 for another twenty years, until the B1 and later the B-2 arrived.

The XB-70 and B-58 did not survive for a number of factors. They cost too much, could not be maintained easily, and could not be easily adapted as the roles for bombers changed. In addition, they emerged at a time when aircraft were being dis-placed in the nuclear weapons delivery role by ballistic missiles. Optimized for high altitude and high speed and little else, they could not compete with the ICBMs being produced in the 1960s. In this contest, they became both redundant, and far too costly. The B-52 survived because of its inherent flexibility, developing specialties never considered by their designers. Beginning in 1965, B-52s began dropping con-ventional bombs in the Vietnam War. They flew in World War II bomber formations, attacking not just industrial targets, but troop concentrations, sometimes in close air support of ground attacks. Significantly, their versatility allowed them to participate in alternative roles beyond those in which they competed with ICBMs.

Fighters

Fighter aircraft after World War II lacked the precedence for development en-joyed by the bombers. Instead, jet fighter development had to rely on the momentum of research on production models and prototypes that emerged immediately at the war's end. As a result, the initial jet fighters designed at the end of World War II had improved performance but were not a radically different from their piston-engine predecessors. Most of these models became operational fighters, in the late 1940s and early 1950s. Of this group, the straight-wing Republic P-84, Lockheed P-80, and Grumman F-9F could fly up to 100 mph faster than their predecessors, but only slightly higher, and these first-generation jets provided only marginal improvements in their assigned roles. Only the P-86 with swept-back wings, so modified after stud-ies of German test results and research, became an exceptional performer. And, it was not until the Korean War brought increases in defense budgets and a pronounced need for fighter aircraft that the pace of fighter development picked up. When the designs for new U.S. jet fighters did emerge in the early 1950s, they showed consid-erable differences from the fighters of World War II, not only in design and propul-sion, but also in their specialized roles.

Where the P-47, P-51, F-6F, and other World War II fighters had served in both air-to-air and air-to-ground roles, postwar fighters tended to become more narrowly specialized in one particular role.[28] A major reason for the divergence from multi-role to single-role fighters came from the previously mentioned performance characteristics of jet engines. A piston-engine had its optimum efficiency at low altitude, the operating regime of the fighter-bomber, whereas jet engines, with their higher fuel consumption at low altitudes, operated at a disadvantage there. Thus, although a piston-engine fighter interceptor designed for operations at high level could adapt easily to low-level operations, a jet fighter had not the same versatility. The poor acceleration and fuel consumption of these first generation of jet engines made these differences more pronounced. As a result, although the early jet fighters mirrored the multi-roles of their predecessors, their limitations soon became apparent, and aircraft designs began to aim at specific roles for each aircraft. Thus began a debate concerning fighter designs that has yet to play itself out: should aircraft be designed and optimized for single role or made more versatile? Is it preferable to develop a force of fewer high-technology fighters for multiple uses or build more lower-technology aircraft for single roles? A case in point: was the A-10, an aircraft designed for ground attack and almost no air-to-air combat capability, a wise investment? The questions, and the debate surrounding them, continued through the post–World War II history of fighter aviation. The subject is mentioned here simply to indicate the origins of the debate. Further discussion of the topic is well beyond the scope of this paper.

With the introductions of jets, fighter aircraft came to assume one or more of four main roles, and, increasingly, aircraft designs sought to optimize for one role, often at great expense to the aircraft's versatility for other uses. The four roles followed from World War II experience: *Fighter escort* of bomber formations, the need for fighters to escort bombers having been learned through bloody experience over Germany (great range a prime requirement to match the long range of the new bombers); *Air-to-air combat,* seen as the basic role for winning command of the air (speed, maneuverability, and armament optimized against employment against similar classes of jet fighters); *Airborne intercept* of enemy bombers, a role involving air-to-air combat, but now with aircraft optimized for intercepting bombers as far from their targets as possible instead of engaging in classic dogfights with other fighters (speed and range optimized); and, *Air-to-ground operations* (so-called fighter bombers) in support of ground forces or surface naval forces, a role exploited with great success in World War II, in both the European and Pacific theaters (with bombing accuracy, loiter time, and weapons capacity optimized). Fighter aircraft designs varied considerably in their utility for more than one of these roles.

The P(later F)-86 became the first widely recognized superior performer of early jet aircraft, and it served in more than one role. Mounting the 6,000-pound-plus thrust J-47 engine (the same engine as in the B-47) compared to the 4,000-pound thrust engines of the other fighter aircraft, it exceeded the speed of sound in 1948 (in a dive). With a capability of carrying bombs, rockets, and with six machine guns, it could fly either as an interceptor or fighter-bomber.[29] This design and subsequent models of the F-86 (with radar to make it an all-weather interceptor) remained on active service in the United States into the 1960s, and many countries around the world into the 1980s.[30]

One further design of jet fighter, the Northrop P-89, began with a more specific role in mind. The P-89 began as an all-weather fighter-bomber designed to overcome deficiencies encountered during the Battle of the Bulge.[31] Soon afterwards, however, the aircraft became a two-seat, two-engine all-weather interceptor. With its straightwings, it neither had the speed of the P-86, nor its maneuverability. The Air Force scrapped plans for it to carry bombs and instead armed it with weapons pods of guns and rockets for its role as an interceptor of bombers. Too slow and sluggish for dogfighting and with no air-to-ground capability, the P (later F)-89 became the first of a long line of role-specialized jet fighters.[32]

The fighter escort role for bombers had the shortest history and can be dealt with rather briefly. The Air Force selected the Republic F-84 to serve in bomber escort duty, following in the path of the Republic P-47 (along with P-51s) that had served this role escorting B-17s and B-24s in Europe during World War II. Just like the P-47, the F-84 also had excellent ground-attack capabilities, a second role it performed.[33] Escorting bombers made great demands on a fighter's range in order to keep pace with the ever-increasing range of bombers that by the 1950s had reached nearly 10,000 miles unrefueled. To equip the F-84 for increased range, it had an aerial refueling wing receptacle installed to permit it to refuel from the same tanker aircraft as those used by the B-47 and B-52. With this capability, the F-84 anticipated by nearly a decade the development of aerial refueled fighters. Aerial refueling became routine for bombers beginning in the early 1950s, but not fighter aircraft until the 1960s (except for fighter deployments overseas).

F-84s continued as fighter escorts until 1956, when that role ended, at least temporarily, amidst a series of shifts in plans. In 1952, the Air Force had directed an expanded employment of escort fighters beginning in 1956. That expansion came based on a recognition of the extended-range aerial refueling provided to the fighters and the improved designs of nuclear weapons (smaller). As a result, the Air Force decided to make the fighters themselves part of the bombing force. Directives stated that in the 1956-60 period, fighter aircraft accompanying bombers "shall be considered as part of the strategic striking forces for the strategic air offensive, not as fighter escorts, even though they may possess a capability for minimizing enemy air defenses in addition to their primary capability as strategic strike aircraft."[34] So certain were the plans that an aircraft to replace the F-84 in 1956 had already been identified, the supersonic, twin-engine F-101.

By 1956, however, the Air Force had essentially reversed its position. With the introduction of the B-52, bomber tactics shifted to having each nuclear weapon's carrying bomber proceed independently to its target, making fighter escort by the F-84 or any other fighter an obsolete concept. The shift in plans reflected resource priorities: by 1956, SAC had opted to have additional B-52s as a higher priority than the six wings of escort fighters it possessed. In the spring of 1957, SAC inactivated or transferred to Tactical Air Command its six fighter wings.[35] It was not until 1972 when B-52s engaged in conventional bombing over North Vietnam in the Linebacker Operations that the escort role for bombers returned.

The first combat test of jet fighter capabilities came in the Korean War, and inevitably there were some surprises. By 1950, though much experimentation had taken place, the operational fighter capabilities of the Air Force and Navy had progressed little since the end of World War II, with the exception of the F-86. Most attention had gone to modifying or otherwise improving the existing designs. As a result, the Air Force F-80 and Navy F-9F, straight-wing, subsonic aircraft, became the first jets to deploy to Korea. These aircraft initially encountered no serious threats to their air supremacy from the North Korean Air Force, since even World War II-era F-51 Mustangs and B-29s could operate successfully in that low-threat environment. That situation lasted until October 1950, when the Soviet-built, swept-wing MiG-15s showed up over North Korea, creating an entirely new threat, and one that neither F-80, F-9F, nor any other straight-wing fighter could deal with successfully.

The Air Force fighter that could match the MiG-15, the F-86, had been retained in the United States for an air defense role. Withholding the F-86 reflected the perception at the time in which the United States had seen the Korean War as a possible opening to a more general Soviet attack as NSC-68 had described, and thus priority went to keeping its frontline fighter home to defend against a possible Soviet bomber attack on the United States. With the MiG-15 threatening U.S. air power in a current hot war, however, the United States dispatched part of the F-86 force to join the war in Korea. For the remainder of the war, the fighter duels between F-86s and MiG-15s in "MiG Alley" in the northwest corner of North Korea became the most prominent aspect of the air war, with the F-86s downing 810 enemy aircraft (including 792 MiGs) while losing only 78 aircraft to MiGs.[36]

While F-86s and the air-to-air combat role received much of the attention and positive reviews, that role comprised only a part of jet fighters' performance in the war; a much larger portion involved interdiction and close air support operations. In these roles, jet fighter bombers received only mixed reviews. Two drawbacks became evident immediately: first, jet aircraft proved less capable than their predecessors in operating from rough runways as found in Korea; second, the relatively short range of the jet fighters—when compared with their piston-engine equivalents—provided far less loiter time for them on station and available to attack targets. Range capability had become particularly important when during the crucial early stages of the war aircraft had to operate from Japanese bases. Capabilities aside, early in the war, F-51 Mustangs had to take up the slack caused by shortages of F-80s, even when that involved taking F-51s out of storage.[37]

Apportioning the high-performance jet fighters to the air-to-air combat role and piston aircraft to ground attack began a practice of single-role identification that continued long after the war. Many considered that the speed of the jets made them less accurate bombers than the slower, lower flying, piston aircraft, even if the jets' speed made them much less vulnerable to ground fire. Navy practice, for instance, was to use their jets in the role of flak suppression, while the Corsairs and Skyraiders carried out the primary strikes against other ground targets.[38] The Air Force maintained that jet aircraft could perform both roles, air-to-air and air-to-ground combat, in a superior manner and continued to employ its F-80s and F-84s in ground attack, but doubts of their relative effectiveness persisted. With the ground attack roles in the Korean War falling extensively to F-51s, Navy and Marine Corps F-4U Corsairs and AD Skyraiders, a significant body of opinion developed, particularly among Army personnel, that emphasized the advantages of using slower piston aircraft over jet aircraft in conducting close air support operations. Based on the experience in that war, the Army called for the development of an aircraft specifically designed for ground attack, and, not coincidentally, for control of the employment of those aircraft by Army, not Air Force, commanders.[39] The differences of opinion between the Army and the Air Force of fighter aircraft most suited for close air support endured long after the war, into the Vietnam War, and beyond. And, the Air Force's preoccupation with speed and its seeming neglect of close air support requirements, and of conventional bombing in general, in this period further exacerbated the problem.

Even if slower piston aircraft had proved more accurate in the Korean War, the growing threats of ground defenses to aircraft survivability favored the employment of jet aircraft. The faster, higher flying jets suffered fewer losses to enemy anti-aircraft artillery and other ground fire, the big killer of aircraft in the war. In the Korean War, the quality of air defenses faced allowed discussions of trade-offs between accuracy and vulnerability. In later wars, the appearance of surface-to-air missiles and integrated air defenses made the environment so deadly that only high-performance jets could survive, and then only by employing electronic countermeasure tactics. During the Vietnam War, high-threat conditions existed over North Vietnam, but not over the South. As a result, those arguing for trade-offs could cite examples from air operations in South Vietnam to bolster their case.

During and after the Korean War, the United States moved rapidly to develop anew generation of jet fighters, air interceptors for defense of the United States, and a follow-on generation of fighter bombers and air-to-air fighters for more general use. The result was what became known as the "Century-Series" aircraft of the Air Force and aircraft of like-capability by the Navy. Interceptors of Soviet long-range bombers achieved a high priority because of the perceived threat of a surprise nuclear strike. For this role, the subsonic aircraft such as the F-86, F-89, and F-94 (derivative of the F-80) did not suffice. In 1951, the Air Force awarded Convair Aircraft a contract for an aircraft that would become the supersonic F-102, a delta-wing aircraft, armed with missiles, but no guns, designed for the very specific role of intercepting Soviet bombers. The aircraft came into the Air Force inventory in 1956, but lasted only five years in the active force before going to the Air National Guard. Convair copied the delta wing for its design of the B-58, and for an improved version of the F-102, the Mach 2 F-106.[40] Also on an interim basis, the Air Force employed the F-104, a Mach 2 aircraft originally developed as a day fighter (that is, it had no radar) successor to the F-86 as an interceptor while awaiting the F-102 to become operational.

In the same time period, the state of the art of air-to-air missile technology had advanced to a degree that seemingly had spelled an end to aircraft dogfights, the close air-to-air combat that had characterized fighter engagements since the birth of combat aviation in World War I. Instead, missile weapons threatened the possible destruction of aircraft from a range of several miles, well outside of the range of an aircraft's guns. If these projections were proved correct, the implications for fighter aircraft design were enormous: specifically, fighters would not need close-in maneuverability, and missiles could replace guns on the fighters. Designers of the F-102 and its successors had planned for this type of employment and produced aircraft optimized for speed of intercept of bomber formation, in the process making them far less versatile for accomplishing other roles.

As a final element to deal with U.S. air defense requirements of the 1950s, the Air Force needed a long-range interceptor to complement the shorter range F-102 and F-106. They accepted the F-101, the aircraft originally developed as an escort fighter for SAC but disposed of by that command after 1957.[41] The F-101 was to serve only as an interim interceptor until the Air Force could secure a Mach 3 aircraft, planned as the F-108. The XF-108 suffered a similar fate to another Mach 3 aircraft, the XB-70. The XF-108 and XB-70, both made by North American Aviation, used in common engines and several other systems, so the cost overruns by one often occurred to the other.[42] In addition, these aircraft also encountered a change in the Soviet threat that basically eclipsed their future development. The development of ICBMs in the Soviet and U.S. arsenals meant on the one hand that there were now alternatives to expensive bombers, not matter their speed. Similarly, airborne fighter interceptors, no matter how fast, could provide no defense against these missiles. As a result, by the 1960s, aircraft designed primarily for bomber intercepts lost much of their value. The F-106 continued in the Air Force inventory as its primary interceptor, but the specialized design of a bomber interceptor disappeared.

Two of the Air Force Century Series were designed with both air-to-air and air-to-ground capabilities in mind, but they served mainly in the latter category. The F-100, the first fighter to exceed Mach 1 in level flight, became a direct successor of the F-86. The F-100, with a J-57 engine, the same engine used by the F-102, F-101, B-52, KC-135, and a number of other aircraft, performed in both the air-to-air and air-to-ground roles. Over 1,000 entered the Air Force inventory from 1952 to the early 1960s and became a frontline combat aircraft into the mid-1960s in the air–to-ground role, including in the initial stages of the Vietnam War. (Some F-101s and F-104s were also produced as fighter-bombers.) By the mid-1960s, however, Mach 2 fighters then appearing as well as the development of surface-to-air missiles greatly limited the F-100's effectiveness.

The F-105 became the frontline Air Force fighter-bomber, but its design owed little to Korean War experience and much to the 1950s concentration on nuclear weapons delivery. Since a prime lesson of the Korean War was a resolve to not engage in such conflicts, the tactical lessons were ignored. The F-105 instead filled the Air Force requirement for a supersonic fighter-bomber to deliver nuclear weapons. Optimized for this role, it had far less capability as an air-to-air fighter. In the Vietnam War, the F-105 adapted well to the role of delivering conventional bombs; it could carry 4,000 pounds of bombs and was responsible for 75 percent of the strike sorties over North Vietnam. For attaining air superiority against North Vietnamese MiGs, however, none of the Century Series could help.

Naval aircraft produced after the Korean War reflected better the lessons from that war. First, the Navy produced an aircraft maximized for ground attack, the A (designating attack)-4. The A-4 had roughly the same speed as the F-80, high subsonic, but it had a far greater weapons carrying ability and with twice its range. The A-4 entered operational service in 1956, and a successor, the A-7, appeared a decade later. As a combined air-to-air fighter and fighter-bomber, the Navy produced the AH-1, later called the FH-1, and then designated the F-4A Phantom II. This two-seat, twin-engine fighter became not only the Navy's frontline fleet defense interceptor, but also was adopted by the Air Force in 1963 as the F-4C.[43] The F-4 became the prime frontline fighter of the Vietnam War, used by both Air Force and Navy for providing air cover over North Vietnam and as fighter-bombers employed by both services.

During the Vietnam War, in addition to the F-4, the Air Force adopted another Navy aircraft for ground attack missions, a piston-engine aircraft the Navy had employed in the Korean War, the AD Skyraider, designated the A-1 by the Air Force. The A-1, a throwback in many ways to the fighter bombers of World War II, represented in many ways a reaction to the post–World War II trend for fighters of ever-increasing speed and altitude capability. The A-1 could not perform even minimally as an air-to-air fighter, nor could it operate against an enemy with a sophisticated air defense network. Over South Vietnam, however, its loiter time, weapons load, and delivery accuracy made it the most effective platform available. In the 1970s, the Air Force developed its own ground attack-specialized aircraft, the A-10. This two-engine jet has a top speed no faster than the World War II fighter-bombers, but it provided great advantages for operations in which the A-1 had excelled. After thirty years, that aircraft still continues in the Air Force inventory for these specialized roles.

Many other developments emerging from the Vietnam War experience gave less meaning to separate accounts of bomber and fighter development. Unmanned aircraft, cargo aircraft outfitted as gunships, B-52s employed conventionally in close air support, and fighters used as strategic bombers blurred definitions and differences. Until the end of the Cold War, the nuclear weapon delivery role tended to keep long-range bombers in a category apart, but more and more distinctions of bombers and fighters became less important or meaningful. The fighter/bomber distinction clearly passed the point of no return with the introduction of the F-117—designated a "fighter" but whose sole capability was dropping bombs. Moreover, while for the first twenty years after World War II, bomber and fighter platforms succeeded one another after every five years, for the next forty years, platforms stabilized but changes of a different kind occurred. Though B-52s are still employed and fighters of today fly no faster and no higher than they did forty years ago, internal systems of those aircraft allow them to perform far different roles from those they were initially configured.

Certainly, von Karman's vision of the future was not complete—he could not know that stealth technologies would trump the need for extreme speed in penetrating enemy territory—but his concepts did describe the experience of the following twenty years of fighter and bomber development, and much else. In the Vietnam War, for instance, aerial tankers began routinely to refuel fighter aircraft, a precursor to present air operations in which tankers are standard complements for fighters and bombers, as well as cargo and electronic systems aircraft, and much else. With that capability, the United States can project, in von Karman's words, airborne task force to strike at far distant points and be supplied.

During the Vietnam War, EC-121s, a 1950s radar early-warning aircraft used by the Air Force and Navy, deployed to Vietnam to provide radar early warning and served as communications relay aircraft. Notably, in October 1967, it made history when it guided a fighter to a successful intercept of a MiG-21 over the Gulf of Tonkin, the first time an airborne controller had directed such an attack.[44] That was the first, and the EC-121 became the prototype for systems appearing later on the E-3A and E-2C. These and other like aircraft have made closure on von Karman's prediction of perfect communication between command and individual aircraft, though in even more extensive parameters. Von Karman had described a system of command and control of aircraft through a central ground control of aircraft and targeting. Computers, sensors, and miniaturization even in the 1960s had begun to move those functions airborne and into unmanned systems, providing operational possibilities we have only begun to realize.

Notes

1. Michael H. Gorn, *Harnessing the Genie: Science and Technology Forecasting for the Air Force, 1944–1986* (Washington, D.C.: Office of Air Force History, 1988), 11–30.

2. Richard Hallion posits six generations by 1974; others name four. Either way the progress was extraordinary, and expensive. See Richard P. Hallion, "A Troubling Past: Air Force Fighter Acquisition Since 1945," *Air Power Journal* 4 (Winter 1990).

3. Richard G. Davis, *Carl Spaatz and the Air War in Europe* (Washington, D.C.: Smithsonian Institution Press, 1992), 539–40.

4. Walter J. Boyne, ed., *Air Warfare, An International Encyclopedia* (Santa Barbara, Calif.: ABC-CLIO, 2002), 262.

5. Wesley Frank Craven and James Lea Cate, eds., *The Army Air Forces in World War II*, Vol. 6, *Men and Planes* (Washington D.C.: Office of Air Force History, 1983), 250–51.

6. Richard P. Hallion, "The Air Force and the Supersonic Breakthrough," in Jacob Neufeld, George M. Watson, Jr. and David Chenoweth, eds., *Technology and the Air Force: A Retrospective Assessment* (Washington, D.C.: Air Force History and Museums Program, 1997), 49.

7. Ibid., 55.

8. Mark D. Mandeles, *The Development of the B-52 and Jet Propulsion: A Case Study in Organizational Innovation* (Maxwell AFB, Ala.: Air University Press, 1998), 60.

9. Michael E. Brown, *Flying Blind: The Politics of the U.S. Strategic Bomber Program* (Ithaca, N.Y.: Cornell University Press, 1992), 124–25.

10. John W. R. Taylor, ed. and comp., *Combat Aircraft of the World* (New York: Putnam's, 1969), 465–67.

11. Marcelle S. Knaack, *Encyclopedia of U.S. Air Force Aircraft and Missile Systems*, Vol. II, *Post–World War II Bombers 1945–1973* (Washington, D.C.: Office of Air Force History, 1988), 3–22; Minutes of the 7th Meeting of the Senior Officers Board, 17 Apr. 1950, Presentation of General Sweeney of Strategic Air Command, Section V, Tab Ia, folder "334; Aircraft and Weapons

Board, 1952," box 57, entry 377A (Deputy Chief of Staff, Material, Executive Office, classified Decimal File, 1951), RG 341 (Headquarters USAF), National Archives (NA), College Park, Md. (Hereafter cited as 7th Meeting, Senior Officers Board)

12. Taylor, *Combat Aircraft,* 465–67.

13. *Jane's Encyclopedia of Aviation* (New York: Portland House, 1989), 642.

14. Taylor, *Combat Aircraft,* 466.

15. LeMay to Chief of Staff, USAF, 19 Feb. 1949, Subject: Range Extension, folder "Putt," box 58, Papers of Curtis E. LeMay, Library of Congress, Washington D.C.; Minutes of the 7th Meeting, Senior Officers Board, 17 Apr. 1950.

16. NSC-68, "A Report to the President Pursuant to the President's Directive of January 31, 1950," enclosure in S. Nelson Drew, ed., *NSC-68, Forging the Strategy of Containment* (Washington, D.C.: National Defense University, 1994), 51–53.

17. The *Question Mark* was the name of a Fokker tri-motor aircraft flown by an Army Air Service crew that stayed aloft for over six days in 1929 while being refueled by other aircraft through a hose arrangement. Done as much as a stunt as to demonstrate a capability, the event took on significance because its crew included future Army Air Force Gens. Carl Spaatz, Ira Eaker, and Elwood Quesada. No significant follow-up occurred to exploit this capability until after World War II. See Thomas A. Julian, "The Origins of Air Refueling in the United States Air Force," in Jacob Neufeld, George M. Watson, Jr., and David Chenoweth, eds., *Technology and the Air Force: A Retrospective Assessment* (Washington, D.C.: Air Force History and Museums Program, 1997).

18. Gen. Sweeney briefing, 7th Meeting, Senior Officers Board, 17 Apr. 1950.

19. *Jane's Encyclopedia of Aviation,* 185

20. Ibid., 456.

21. Brown, *Flying Blind,* 101–3.

22. Mandeles, *The Development of the B-52,* 135–44.

23. "B-49 Flying Wing," in Global Security.org, website: <http://www.globalsecurity.org/wmd/systems/b-49.htm>, accessed 10 Sept. 2003.

24. Air Force Link, "B-52 Stratofortress," website: <http://www.af.mil/factsheet.asp?id=83>, accessed 22 Sept. 2006.

25. Brown, *Flying Blind,* 169, 234.

26. Ibid., 185–86; Knaack, *Post–World War II Bombers,* 353–83.

27. Brown, *Flying Blind,* 219–29.

28. On the subject of overspecialization, see Hallion, "Troubling Past."

29. Charles T. Kamps, "The F-86 Sabre," *Air Power Journal* 17 (Summer 2003).

30. Taylor, *Combat Aircraft,* 540.

31. "Presentation to Deputies Council," 30 Apr. 1951, folder "Status of aircraft," entry 10 (Office of the Chief of Staff, Special Assistant to the Chief of Staff, Decimal file), RG 341 (Headquarters USAF), NA.

32. The Air Force changed aircraft designations in June 1948, with F (fighter) replacing P (pursuit). For the remainder of the paper, I will shift to the now more well known "F" designation.

33. Taylor, *Combat Aircraft,* 551–52.

34. Memorandum, Deputy Chief of Staff, Operations (Lt. Gen. Thomas White), for all Deputies, Subject: Concept for the Employment of Fighter Aircraft in the Strategic Air Forces, 8 Sept. 1952, folder "Aircraft, 5," box 34, Papers of Hoyt S. Vandenberg, Library of Congress.

35. Robert Frank Futrell, *Ideas, Concepts, Doctrine: Basic Thinking in the United States Air Force, 1907–1960* (Maxwell AFB, Ala.: Air University Press, 1989), 512.

36. Wayne Thompson and Bernard C. Nalte, *Within Limits: The U.S. Air Force and the Korean War* (Washington, D.C.: Air Force History and Museums Program, 1996), 27–28.

37. Conrad Crane, *American Airpower Strategy in Korea, 1950–1953* (Lawrence: University of Kansas Press, 2000), 24–25; Victor Flintham, *Air Wars and Aircraft* (New York: Facts on File, 1990), 220–21, 224.

38. Crane, *American Airpower,* 111, 135, 177.

39. John Schlight, *Help from Above: Air Force Close Air Support of the Army, 1946–1973* (Washington, D.C.: Air Force History and Museums Program, 2003), 154–59.

40. Taylor, *Combat Aircraft,* 466.

41. Ibid., 523–24.

42. Mark A. Lorell and Hugh P. Levaux, *The Cutting Edge: A Half Century of U.S. Fighter Aircraft R&D* (Santa Monica, Calif.: Rand, 1998), 70–71.

43. The decision to adopt a Navy aircraft was one made not by the Air Force but by the Secretary of Defense Robert McNamara, based on studies conducted by his Systems Analysis Office. The history of this decision as well as discussion of the decision to produce a single common fighter aircraft for the Air Force and Navy, the TFX, is beyond the scope of this paper. See Hallion, "Troubling Past," for more on this topic.

44. "EC-121 Warning Star," in Federation of American Scientists, website: <fas.org/man/dod-101/sys/as/ec-121.htm>, accessed 10 Sept. 2003.

General Bernard Schriever:
"Father of Air Force Missiles and Space Power"

Jacob Neufeld

In 2002, the National Aeronautics and Space Administration (NASA) published a pamphlet entitled "Celebrating a Century of Flight." It reads in part under the heading "Dawn of the Space Age":

> The National Aeronautics and Space Administration (NASA) was formed with President Dwight D. Eisenhower's signing of the National Aeronautics and Space Act of 1958. The NACA [National Advisory Committee on Aeronautics] and parts of other agencies formed its core; its purpose was research and development for the exploration of space. NASA also emerged in some measure because of the pressures of national defense during the Cold War with the Soviet Union, a broad contest over ideologies and allegiances of the nonaligned nations of the world in which space exploration emerged as a major area of contest.
>
> The space age actually began before the creation of NASA, from the late 1940s, the Department of Defense pursued research, rocketry, and upper atmospheric sciences as a means of ensuring American leadership in technology. A major step forward came when President Eisenhower approved a plan to orbit a scientific satellite as part of the International Geophysical Year (IGY), a cooperative effort to gather scientific data about Earth, for an eighteen-month period from 1957 to 1958. The Soviet Union quickly followed suit, announcing plans to orbit its own satellite.
>
> The Naval Research Laboratory's Project Vanguard was chosen on 9 September 1955 to support the IGY effort, largely because it did not interfere with high-priority ballistic missile programs, while an Army proposal to use the Redstone ballistic missile as the launch vehicle waited in the wings. The technological demands upon the Vanguard program were too great and the funding levels too small to ensure success.[1]

What we have here is revisionist history at its worst. While acknowledging the fact that the Space Age predated its creation, NASA fails to even mention the military's premier ballistic missiles development program—the one led by Air Force General Bernard A. Schriever. It makes no mention of the United States Air Force—as if NASA, the Army, and the Navy were the only ones responsible for and contributed to U.S. space program. The purpose of my essay is to correct this pamphlet's distortions and present the true history of American military missile research and development.

The Real Story

On 19 February 1957, in a speech entitled "ICBM—A Step toward Space Conquest" delivered at a symposium on space flight in San Diego, Commander of the Western Development Division Major General Bernard A. Schriever stated:

Our safety as a nation may depend upon our achieving "space superiority." Several decades from now the important battles may not be sea battles or air battles, but space battles, and we should be spending a certain fraction of our national resources to insure that we do not lag in obtaining space supremacy. Besides the direct military importance of space, our prestige as world leaders might well dictate that we undertake lunar expeditions and even interplanetary flight.

Thus, even as he labored to develop the intercontinental ballistic missile (ICBM), Schriever sensed the importance of space. He recognized that the same boosters that propelled the ICBM could also put the United States on the threshold of space. Foreseeing the importance and urgency to be the first nation to enter space, Schriever wanted his Division to assume responsibility for space research and development (R&D). His aim was to demonstrate that, because of the great progress being made in developing the ICBM, the nation had a unique opportunity to invest in the militarily important space program.[2]

Secretary of Defense Charles Wilson, however, discerned that Schriever's speech was contrary to the administration's policy, which promoted the "peaceful uses of space." Consequently, Wilson ordered the Air Force to order Schriever to stop using the word "space" in his future speeches.

Let us flash back to 1946. A protege of General Henry H. "Hap" Arnold, Colonel Schriever headed the Air Staff's Scientific Liaison Office in the Pentagon. His assignment was to ensure that the successful partnership between airmen and scientists, forged during World War II, would continue. Moreover, Arnold asked Air Staff planners to look ahead fifty years—into the future. Arnold was absorbed with the need to avoid a nuclear Pearl Harbor and believed that the United States could avoid such a catastrophe provided that the nation obtained the best available form of strategic intelligence. Was an Earth satellite feasible, he wondered?

The answer came on 2 May 1946 when the RAND Corporation published its landmark study, called "Preliminary Design of an Experimental World-Circling Spaceship," that suggested it was feasible to launch satellites into space by using existing multi-stage rockets. In fact, however, this feasibility had already been demonstrated back in October 1942 when a German V-2 test rocket had scraped the edge of space during a test flight over the Baltic Sea.

Also in 1946, Dr. Louis Ridenour, the Air Force's first chief scientist, identified all of the Air Force's likely space missions: strategic reconnaissance, surveillance, weather, communications, and navigation. But there were some prominent skeptics who consigned missiles and space to the "Buck Rogers" comic books. For example, Dr. Vannevar Bush, wartime head of the Research and Development Board, ridiculed Arnold's approach as infeasible and unimportant.[3]

Subsequently, the RAND study led to the Project Feed Back reports of the 1950s, which advocated promoting space R&D. Nevertheless, nearly a decade went by before the Air Force embarked on a viable space program. The impetus behind this decision was the 1953 thermonuclear breakthrough. Even then, the Air Force took a conservative approach. It issued a General Operational Requirement for a strategic satellite reconnaissance system. Issued in March 1954, a RAND report by James Lipp and Robert Salter predicted that such a satellite would not be available for ten years.

The most urgent R&D program for the Air Force during the 1950s was the development of an ICBM, intercontinental ballistic missile. In August 1954, after overcoming inter- and intra-service rivalries, the Air Force established the Western Development Division and assigned the top job to Brigadier General Schriever. Anxious to counter the Soviet military threat, President Eisenhower assigned the top national priority to Schriever's ICBM program. Then, as the Cold War heated up, several defense studies recommended deploying intermediate-range ballistic missiles (IRBMs) as an interim solution, and developing the U-2 aircraft for collecting reconnaissance information over the Iron Curtain.

The Air Force's Advanced Reconnaissance (Satellite) System (WS-117L) was a long-term program, projected to cost more than $100 million and would not become operational until 1963. However, only $4 million was allotted to the effort and it was limited to studies and testing.[4]

A longtime advocate of "technology push," that is, allowing technological development to progress to wherever it led, General Schriever supported pursuing both the IRBM and the space satellite programs in addition to the ICBMs. However, in order to make certain that these efforts would not drain off scarce scientific talent and resources from the ICBM, Schriever advocated incorporating both efforts under his Division. Subsequently, in the autumn of 1955, he requested to transfer management of the WS-117L satellite from Wright-Patterson Air Force Base's Wright Air Development Center, in Ohio, to his Western Development Division, in Los Angeles. The transfer was carried out in February 1956. The Air Force also obtained Camp Cooke, California, for its space launches. The facility was later renamed Vandenberg Air Force Base.[5]

The Air Force's military satellite system was designed to perform several separate missions, including photoreconnaissance and missile warning. By 1959, it had evolved into three major elements: the Discoverer program, the Satellite and Missile Observation System (SAMOS), and the Missile Detection and Alarm System (MIDAS). Discoverer and SAMOS conducted photoreconnaissance, and MIDAS performed missile warning.

Other R&D satellites evolved to perform various support missions, such as nuclear surveillance, weather observation, navigation, and communication. In June 1956, Schriever's Western Development Division contracted with Lockheed to provide new space boosters. The program underwent several name changes, from "Pied Piper," to "Hustler," and finally, "Agena."

Because of his emphasis on strategic reconnaissance, President Eisenhower insisted that no single military service control this "vital national asset." At the end of his second term, the president set up the National Reconnaissance Office. Although it came under the intelligence community, it was headed by the Under Secretary of the Air Force.[6]

General Schriever lobbied earnestly to secure financial support for missiles and space systems. He frequently took the "red eye" flight from Los Angeles to Washington to testify before congressional committees. President Eisenhower, however, maintained his policy of "space for peaceful purposes" and limited federal spending. As a result, sizeable reductions were made to the projected size of the ICBM force and limits were set with respect to R&D funds for space programs.

On 4 October 1957 came the electrifying shock of *Sputnik,* with the Soviet Union's orbiting of *Sputnik,* the world's first artificial satellite. Considering that *Sputnik* was launched by an ICBM and that the USSR had achieved a breakthrough in thermonuclear weapons served as a "reality check" that demanded an American response. Clearly, Schriever's earlier call for action on space R&D was now irrefutable. Although some cuts that had been made to the ballistic missiles program were now restored, the administration's "peaceful purposes for space" policy remained in effect.

On the other hand, the Air Force's senior leadership now became more proactive. Chief of Staff General Thomas White, in an address to the National Press Club on 29 November 1957, coined the term "aerospace," arguing that a continuum existed between air and space. The Air Force now billed itself as an aerospace power; the theory and nature of aerial warfare took a quantum leap forward. Planners prepared for combat not only in the atmosphere, but also discussed the potential outer space arena.[7]

General Schriever continued to advocate the Air Force's commitment to space superiority. In a letter to the commander of Air Matériel Command (AMC), Schriever wrote: "superiority in space is a fundamental requirement since on it will depend the entire future position, prestige, and indeed, the welfare of the U.S."[8] Meanwhile, in February 1958, part of the Defense Department's centralization plan provided for the creation of the Advanced Research Projects Agency (ARPA), which would conduct R&D for military space programs. In October, NACA was abolished and NASA was created.

Despite encountering difficulty in obtaining adequate support for a military space program, Schriever's ICBM team achieved several noteworthy milestones. For example, in December 1958 an Atlas launch placed the entire missile into orbit and in February 1959, a Thor-Agena booster launched Discoverer I into polar orbit. Between February 1959 and February 1962, Discoverer—a cover for the CORONA program—carried out thirty-eight launches, including the world's first polar orbiting satellite. Another achievement was the 1960 mid-air recovery of reconnaissance film from space aboard Discoverer XIV, three years better than predicted.

In April 1959, Schriever was named to head the Air Research and Development Command (ARDC), and promoted to lieutenant general. That September, the military services regained from ARPA the right to conduct space R&D. The Air Force assumed responsibility for reconnaissance and surveillance satellites and for developing and launching all military space boosters; the Army managed communications satellites; and the Navy oversaw navigation satellites. As a three-star general, Schriever continued to advocate military space power. In a December 1959 speech, he emphasized that although spectacular feats in space might enhance a nation's prestige, it was not his major concern: "My really pressing concern is the direct and immediate importance of exploiting the advantage that space offers to our vital military deterrent posture."[9]

By 1960, the Air Force was able to point to several tangible accomplishments in space: its Atlas and Thor missiles provided most of the space launch boosters and much of the infrastructure support for NASA. On the military side, the Air Force had readied an early-warning satellite and operated a ground-based surveillance system.

The Kennedy Administration

The inauguration of the John F. Kennedy administration ushered in a new urgency for military space systems. The Kennedy administration emphasized three interrelated space activities: military satellites, manned spacecraft with military uses, and making greater use of missiles as satellite boosters.[10]

Two important studies set the stage. Schriever knew both authors well. First came Jerome B. Wiesner of MIT's report of 10 January 1961 which called for military participation in the Moon landing. Then, in March, Trevor Gardner (former Assistant Secretary of the Air Force for R&D) completed a space study which concluded that without a significant infusion of funding, the United States would not overtake the Soviet Union in the space arena for up to five years.[11]

Deputy Defense Secretary Roswell Gilpatric—another Schriever associate and formerly the under secretary of the Air Force and Aerospace Corporation executive—proposed assigning space R&D to the Air Force, provided the service was able to "get its house in order." This meant patching up the confused organizational arrangement between Air Force Research and Development Command, ARDC, and AMC—where ARDC was responsible for R&D, while AMC controlled the funding. Chief of Staff General Thomas White had to step in to resolve the problem.[12]

Subsequently on 6 March 1961, Defense Secretary Robert McNamara issued Directive 160.32, which assigned space R&D to the Air Force—making the service the executive agent for space. Also, effective on 1 April, the Air Force Systems Command and Air Force Logistics Command were created and Schriever was named to head AFSC.

Anti-satellite Program

In February 1963, the Air Force approved development of a ground-based anti-satellite system. Designated Program 437, it employed Thor missiles with nuclear warheads to destroy or disable hostile satellites or space-based weapons. Thor missiles, taken from deactivated sites in the United Kingdom, were stationed on Johnston Island in the Pacific. Three of four successful tests were launched in May 1964 and the system was declared operational under Air Defense Command.

The anti-satellite capability remained in place until the mid-1970s. In 1975, the Air Force's Space and Missile Systems Organization undertook development of nonnuclear systems. The most promising program emerged several years later as a two-stage miniature homing vehicle carried aloft by an F-15 aircraft. In March 1988, however, congressional restrictions and budgetary considerations led to its termination.

Manned Space Flight

General Schriever supported military manned space flight and sought to extend the Air Force's traditional missions into space: "surveillance and reconnaissance, interception, bombardment, and command, control, and communications." However, neither Eisenhower nor Kennedy administration officials had any interest in moving the arms race into space.[13]

While the most attention was lavished on unmanned vehicles, Schriever also expressed interest in manned space systems. Thus, in December 1963, Secretary McNamara cancelled the Dyna-Soar (X-20) project, the Air Force's program for manned missions in the upper atmosphere and near-earth orbit. McNamara had cancelled the X-20 primarily because it would have accelerated the arms race, and secondly because President Lyndon Johnson believed that NASA's Gemini could do the job better. Yet, McNamara approved the Air Force's Manned Orbiting Laboratory (MOL), a program for manned military space research. The irony was that the Air Force was assigned a technology research job, while NASA received an operational assignment.

General Schriever placed the MOL program under his Space Systems Division. The MOL's only launch occurred in November 1966. Among its noteworthy successes, the MOL's heat shield survived a sub-orbital reentry. Nonetheless, it too was cancelled. The Air Force–produced Titan IIIC launch vehicle (based on the Titan II ICBM) was intended to have been mated to NASA's Gemini spacecraft, to lift off the MOL. In June 1969, a lack of funds caused the demise of the MOL, as Apollo progress had rendered Gemini a less attractive option. Another factor was the rising cost of the Vietnam War.

Over the MOL program's lifetime, seventeen men trained as astronauts, including the Air Force's General Robert Herres and the Navy's Admiral Richard Truly. Despite cancellation of the Air Force's manned military space program, the Air Force enjoyed great success with respect to boosters and instrumented satellites for communications, weather, reconnaissance, warning, and navigation.

More Reorganizations

By September 1966, when General Schriever retired, he had been at the center of Air Force R&D for more than two decades. Moreover, his influence has endured for even a longer period. He could take satisfaction in the knowledge that Air Force boosters had provided the lift for many of NASA's launches. And the Air Force had also sent to NASA some of its best managerial talent—for example, General Sam Phillips, who directed Apollo—as well as a fair share of NASA's astronauts.

Together with their associated upper stages, Thor and Atlas ballistic missiles served as the workhorses of the U.S. space program. In 1959, NASA had started using

Atlas and began to develop the Delta upper stage for Thor. The Centaur was an Air Force program until it was transferred to NASA in 1960. The last Thor was launched in July 1980; the last modified Atlas in March 1995. The Atlas was also used as a booster for the Mercury and Gemini manned space programs.

On 8 September 1970, Deputy Defense Secretary David Packard reiterated that development, production, and deployment of space systems for warning and surveillance of nuclear weapons and launchers would remain with the Air Force. However, future developments would fall under the same guidelines as for other major weapons.

By 1980, the USAF space effort had grown larger, more complex, and increasingly fragmented. For example, warning and surveillance functions and systems were assigned to the Strategic Air Command, air defense aircraft went to the Tactical Air Command, while operational control rested with the North American Air Defense Command, and Air Force Systems Command operated many experimental systems, especially space systems.

Air Force Space Command was activated on 1 September 1982, and three years later the U.S. Space Command was created to provide all of the services access to space.

Schriever's Legacy

Having begun this essay with a quotation from General Schriever, I would like to conclude it with another:

> It may be said that warfare has acquired a new phase—technological war. In the past, research and development were only preparation for the final and decisive testing of new weapons in battle. Today the kind and quality of systems which a nation develops can decide the battle in advance and make the final battle a mere formality—or can bypass conflict altogether.[14]

These prophetic words were written over forty years ago. I think that the veterans of our recent war in Iraq will agree that our superior weapons—including space-based support—ensured victory, shortened the war, and saved a great many lives. For this, we can all thank General Bernard A. Schriever, United States Air Force.

Notes

1. Pamphlet, "Celebrating a Century of Flight" (Washington, D.C.: NASA, SP-2002-09-511-HQ, 2002), 18.

2. John L. Frisbee, ed., *Makers of the United States Air Force* (Washington, D.C.: Office of Air Force History, 1987), 298.

3. Bernard A. Schriever, "Military Space Activities: Recollections and Observations," in R. Cargill Hall and Jacob Neufeld, eds., *The U.S. Air Force in Space: 1945 to the Twenty-first Century* (Washington, D.C.: USAF History and Museums Program, 1998), 13.

4. Ibid., 15.

5. Jacob Neufeld, *The Development of Ballistic Missiles in the United States Air Force, 1945–*

1960 (Washington, D.C.: Office of Air Force History, 1990), 144; David N. Spires, *Beyond Horizons: A Half Century of Air Force Space Leadership* (Peterson AFB, Colo.: Air Force Space Command, 1997), *passim*.

6. R. Cargill Hall, "Civil-Military Relations in America's Early Space Program," in Hall and Neufeld, eds., 30.

7. Neufeld, 243.

8. Ibid., 210n71; ltr, BAS, Comdr. WDD to Gen. Sam Anderson, Cmdr. AMC, no subj., 21 May 1958.

9. *New York Times,* 21 Dec. 1959.

10. Bernard C. Nalty, ed., *Winged Shield, Winged Sword: A History of the United States Air Force, Volume II, 1950–1997* (Washington, D.C.: Air Force History and Museums Program, 1997), 225ff.

11. David N. Spires, "The Air Force and Military Space Missions: The Critical Years, 1957–1961," in Hall and Neufeld, eds., 33–45.

12. Nalty, ed., 225ff.

13. Adam L. Gruen, "Manned versus Unmanned Space Systems," in Hall and Neufeld, eds., 70.

14. Bernard A. Schriever, "The Operational Urgency of R&D," *Air University Quarterly Review* (Winter–Spring 1960–61), 230.

Contributors

Tami Davis Biddle is the George C. Marshall Chair of Military Studies in the Department of National Security and Strategy at the U.S. Army War College—the Army's senior-level staff college. She teaches national security strategy and military history. She was the 2001–2002 Harold K. Johnson Visiting Professor of Military History at the U.S. Army's Military History Institute. Prior to that she taught in the Department of History at Duke University (lecturer, 1992–95; assistant professor, 1995–2001), where she was a core faculty member of the Duke–University of North Carolina Joint Program in Military History. She received her Ph.D. from Yale University, training with Paul Kennedy and Gaddis Smith. She has held fellowships and visitorships from Harvard University's Belfer Center for Science and International Affairs, the Social Science Research Council, the Brookings Institution, and the Smithsonian Institution's National Air and Space Museum. Her research focus has been warfare in the twentieth century, especially warfighting and diplomacy during the two world wars, and the early Cold War period. In particular, she has concentrated on the history of air warfare, and the history of the Cold War. She has written many articles and book chapters on these subjects including, recently, "Sifting Dresden's Ashes," *Wilson Quarterly* (Spring 2005). Her book, *Rhetoric and Reality in Air Warfare: The Evolution of British and American Ideas about Strategic Bombing, 1914–1945* (Princeton University Press, 2002), was a *Choice* outstanding academic book, 2002, and was recently added to the Royal Air Force Chief of Air Staff's Reading List.

Walter J. Boyne is a former Director of the National Air and Space Museum of the Smithsonian Institution. A career Air Force officer, Boyne retired as a colonel with 5,000 hours flying time. He is also the former Chairman of the Board of *Wingspan, the Air and Space Aviation Channel,* and President of his own firm, Walter Boyne Associates. The author of forty books, he is one of the few persons to have had best sellers on both the fiction (*The Wild Blue*) and the nonfiction (*Weapons of Desert Storm*) lists of the *New York Times*. He has had extensive television experience, hosting two talk-shows and hosting and narrating video productions of two of his books, *Beyond the Wild Blue* and *Clash of Wings*. He is an honor graduate of the University of California at Berkeley (BSBA), the University of Pittsburgh (MBA), and has an honorary doctorate from Salem University. Walter Boyne is scheduled to be enshrined in the National Aviation Hall of Fame on July 21, 2007.

James S. Corum is currently a Professor at the U.S. Army Command and General Staff College, Ft. Leavenworth, Kansas. He holds a M.A. from Brown, M. Litt. from Oxford, and a Ph.D. from Queen's University, Canada. Corum is the author of four military history books, including *The Roots of Blitzkrieg: Hans von Seeckt and German Military Reform* (1992), *The Luftwaffe: Creating the Operational Air War 1918-1940* (1997), *The Luftwaffe's Way of War: German Air Doctrine 1911-1945* (with Richard Muller, 1998), and *Airpower in Small Wars: Fighting Insurgents and Terrorists* (with Wray Johnson, 2003). His fifth book, on counterinsurgency strategy, is being published by Zenith Press in 2007. He has authored more than forty major book chapters and journal articles on a wide variety of air power and military history subjects. From 1991 to 2004, Corum was a professor at the USAF School of Advanced Air and Space Studies at Maxwell AFB. In 2005, he was a visiting fellow at All Souls College, Oxford University, and was also elected to a fellowship in the Levershulme Programme at Oxford. Corum retired as a lieutenant colonel in the U.S. Army Reserve and served in Iraq in 2004.

Tom D. Crouch is Senior Curator of the Division of Aeronautics at the National Air and Space Museum, and has been a Smithsonian employee since 1974. He received his Ph.D. in history from the Ohio State University (1976), and was awarded the degree of Doctor of Humane Letters from the Wright State University in 2001. Crouch is the author or editor of more than a dozen books and many articles for both popular magazines and scholarly journals. Most of his work has been on aspects of the history of flight technology. His leading books include: *The Bishop's Boys: A Life of Wilbur and Orville Wright* (1989); *Wings: A History of Aviation from Kites to the Space Age* (2003); *Eagle Aloft: Two Centuries of the Balloon in America* (1983); and *A Dream of Wings: Americans and the Airplane, 1875-1905* (1981). He is the recipient of a number of major writing awards, including the history book prizes offered by both the American Institute of Aeronautics and Astronautics and the Aviation/Space Writers Association. He received a 1989 Christopher Award, a literary prize recognizing "significant artistic achievement in support of the highest values of the human spirit," for *The Bishop's Boys*. Throughout his career, Crouch has played a major role in planning museum exhibitions for a number of institutions, including the Neil Armstrong Museum (Wapakoneta, Ohio); the Ohio Historical Center (Columbus, Ohio); and both the National Air and Space Museum and the National Museum of American History. He is particularly proud of having served as curator for the exhibition, "A More Perfect Union: Japanese Americas and the U.S. Constitution," which opened at the National Museum of American History in October 1985, and remained in place for over fifteen years. In the fall of 2000, President Clinton appointed Crouch to the Chairmanship of the First Flight Centennial Federal Advisory Board, an organization created to advise the Centennial of Flight Commission on activities planned to commemorate the 100th anniversary of powered flight.

Dik Alan Daso (B.S., USAF Academy 1981; M.A., Ph.D., University of South Carolina) is Curator of Modern Military Aircraft at the Smithsonian Institution, National Air and Space Museum. Additionally, he was a co-curator for *The Price of Freedom: Americans at War,* a permanent exhibition at the National Museum of American History that opened in November 2004. A retired Air Force lieutenant colonel, he has served as an RF-4C Phantom instructor pilot, F-15 Eagle pilot, twice as a T-38 Talon instructor pilot, instructor of history at the USAF Academy, and Chief of Air Force Doctrine at Headquarters Air Force, Pentagon. During his career, Daso accumulated over 2,750 total flying hours. He contributed chapters to *The Air Force,* an illustrated history of that service, and *West Point: Two Centuries and Beyond,* a bicentennial publication for the U.S. Military Academy. Daso has written four books: *Architects of American Air Supremacy: Gen. Hap Arnold and Dr. Theodore von Kármán* (Maxwell AFB, Ala.: Air University Press, 1997); *Hap Arnold and the Evolution of American Airpower* (Washington, D.C.: Smithsonian Institution Press, 2000), winner of the 2001 American Institute of Aeronautics and Astronautics (AIAA) History Manuscript Award; *Doolittle: Aerospace Visionary* (Washington, D.C.: Potomac Books, 2003); and *U.S. Air Force: A Complete History* (New York: Hugh Lauter Levin, 2006).

John F. Guilmartin, Jr. is a Professor of History at The Ohio State University where he teaches military history and early modern European history. He received his Bachelor of Science degree and second lieutenant's commission from the USAF Academy in 1962, and earned his M.A. (1969) and Ph.D. (1971) in history from Princeton University. He has written extensively on military history and the theory of war. His publications include *Gunpowder and Galleys: Changing Technology and Mediterranean Warfare at Sea In the Sixteenth Century* (1974; 2d rev. ed. 2003); *America in Vietnam: The Fifteen Year War* (1991); *A Very Short War: The* Mayaguez *and the Battle of Koh Tang* (1995); and *Galleons and Galleys* (2002). He was the editor of *Encyclopædia Britannica's* article on the technology of war for the 1992 edition and subsequent printings. He was editor-in-chief and principal author of Vol. IV of the U.S. Air Force *Gulf War Air Power Survey Report, Weapons, Tactics and Training* (1993). He is presently writing a history of the Vietnam War for Harvard University Press. Guilmartin retired from the Air Force as a senior pilot and lieutenant colonel in 1983. He served as an instructor with the Department of History at the USAF Academy in 1970–74 and was editor of the *Air University Review,* the professional journal of the Air Force, in 1979–83. His staff assignments included duty as Chief of Tactics, Aerospace Rescue and Recovery Service, during 1977–79. His operational career included two Southeast Asia combat tours as a "Jolly Green" long-range rescue helicopter pilot, flying HH-3Es in 1965–66 and HH-53Cs in 1975. During his first tour, he was credited with 120 combat missions over Laos and North Vietnam. During his second tour, he logged an additional 7 combat missions in the April 1975 evacuations of Phnom Penh and Saigon, operating from the USS *Midway* in the latter operation. His decorations include the Silver Star with one Oak Leaf Cluster, the Legion of Merit, and the Air Medal with five Oak Leaf Clusters.

Thomas A. Keaney is the executive director of the Foreign Policy Institute at the Paul H. Nitze School of Advanced International Studies (SAIS), Johns Hopkins University, Washington D.C. He also serves as the executive director of the Merrill Center for Strategic Studies and senior adjunct professor of strategic studies at SAIS. Until 1998, he was a professor of military strategy at National War College, Washington D.C., and director of its core courses on military thought and strategy. During 1991–92, he was a researcher/author with the Gulf War Air Power Survey. He was co-author of two reports of that survey: *The Summary Report* and *The Effects and Effectiveness of Air Power* (both published by the U.S. Government Printing Office in 1993). He is also author of *Strategic Bombers and Conventional Weapons: Air Power Options* (1983) and (with Eliot A. Cohen) *Revolution in Warfare?: Air Power in the Persian Gulf* (1995). His other publications include (ed. with Barry Rubin) *U.S. Allies in a Changing World* (2000) and *Armed Forces in the Middle East, Politics and Strategy* (2002). He is a graduate of the National War College. He holds a B.S. from the U.S. Air Force Academy and M.A. and Ph.D. degrees in history from the University of Michigan. During a career in the U.S. Air Force, he served in positions including: associate professor of history at the U.S. Air Force Academy; planner on the Air Staff; forward air controller in Vietnam; and B-52 squadron commander. He retired as a colonel in 1991.

Jacob Neufeld has been affiliated with the Air Force History and Museums Program for nearly forty years as a staff historian, branch chief, division chief, senior historian, and currently is the Director of the Air Force Historical Studies Office in Washington, D.C. He is also the editor of the journal *Air Power History.* Neufeld earned B.A. and M.A. degrees from New York University and did doctoral studies at the University of Massachusetts, Amherst. Commissioned in the U.S. Army through R.O.T.C., he served with the Corps of Engineers from 1964 to 1966. He has written and edited numerous works in military history and the history of technology, including *The Development of Ballistic Missiles in the United States Air*

Force, 1945-1960 (1990). For twenty years he was an adjunct professor at the University of Maryland, Montgomery College, and American Military University. He is writing a biography of General Bernard Schriever.

Merle L. Pribbenow II joined the Central Intelligence Agency in 1968 after graduating from the University of Washington with a B.A. degree. He served in the CIA for twenty-seven years, including five years in Vietnam (1970–75), as an operations officer and Vietnamese-language specialist. He retired from the Agency in 1995. Since his retirement, Mr. Pribbenow has devoted his newly found free time to research and writing on the Vietnam War and to translating Vietnamese-language publications and documents on the war. His translation of the official Vietnamese Communist history of the Vietnam War, *Victory in Vietnam: The Official History of the People's Army of Vietnam, 1954–1975,* was published by University Press of Kansas in 2002. He has also written a number of published articles on various aspects of the Vietnam War.

Vance R. Skarstedt, Lieutenant Colonel, USAF, graduated from the United States Air Force Academy and holds an M.S. from the University of Southern California and a Ph.D. in history from Florida State University. He served for four years as the Deputy Head of the Department of History at the Air Force Academy and is currently its Acting Head.

Michael Robert Terry is a retired Air Force officer and command pilot with 2,600 flight hours as aircraft commander, instructor, and evaluator pilot. He is a 1974 Air Force Academy graduate and served as deputy director for military history in the Academy's Department of History, where he taught numerous military history courses. He also taught military history for the University of North Dakota. The author of numerous articles for *Air Power History* and *Quest Quarterly,* he is the author of *The Historical Dictionary of the Air Force and its Antecedents,* published by Scarecrow Press. He contributed to the Air Force Chief of Staff's 50th Anniversary volume analyzing the historical development of doctrine and provided a bibliographic essay on air and space power history to the revised Army Center of Military History's *A Guide to the Study and Use of Military History.* He co-chaired the Air Force Academy's Twentieth Military History Symposium—*WINGED CRUSADE: The Quest for American Air and Space Power.* He presently is Deputy for Cadet Heritage, Department of History, USAF Academy and is an Assistant Professor of History where he chairs and teaches numerous courses on air and space power and military history.

Herman S. Wolk recently retired as Senior Historian, U.S. Air Force. After receiving B.A. and M.A. degrees from the American International College, Springfield, Mass., he studied at the Far Eastern and Russian Institute, University of Washington, 1957–59. He was historian at Headquarters, Strategic Air Command, 1959–66, and served in the Office of Air Force History in Washington, D.C. from 1966 to 2005. A fellow of the Inter-University Seminar on Armed Forces and Society, he served on the OSD Project on the Strategic Arms Competition in 1973–74. Wolk is the author of *Strategic Bombing: The American Experience* (1981); *Planning and Organizing the Postwar Air Force, 1943–1947* (1984); *The Struggle for Air Force Independence, 1943–1947* (1997); *Fulcrum of Power: Essays on the Air Force and National Security* (2003); and *Reflections on Air Force Independence* (2007). He is contributing author to *We Shall Return! MacArthur's Commanders and the Defeat of Japan* (1988); *The Pacific War Revisited* (1997); and *Winged Shield, Winged Sword: A History of the United States Air Force* (1997).

Index

A-4, 179
Acceleration research, 46–47
Advanced Reconnaissance System, 185
Advanced Research Projects Agency, Defense Department, 186
Aerial arms race, 35
Aerial deployments, 41–43
Aerial Experiment Association, 16
Aerial refueling, 4, 171, 175, 180
  Arnold's contribution, 39–40
Aero Club de France, 16
Aerodynamic control, 11–12
Aerodynamics, Doolittle's research on, 46–47
Aerodynamic testing device, 13
Aeronautical engineers, German, 88–89
Aeronautical Society of Great Britain, 13
Aeronautic Society of New York, 17
Aeronautics research, 46–47
Aerospace, 186
Aerospace Expeditionary Forces, 34
AH-1, 179
Airborne fighter interceptors, 178
Airborne guided missiles, 164
Airborne intercept, 174
Airborne Warning and Control System, 74
Air Combat Command, 34
Air Corps
  Browning Board report of 1936, 28
  split command, 28
  War Department view, 28
Air Corps Act of 1926, 28
Air Corps Tactical School, 5, 58–59, 65, 85
Aircraft; *see also* World War II aircraft ranking
  construction in 1941, 39–40
  design revolution, 23
  development by Army Air Corps, 142–43
  development time, 141–42
  earliest pilots, 15–16
  engine technology, 16
  European improvements, 16
  first all-metal, 16
  first marketing of, 15
  first water take-off, 16
  with floating wing tips, 169
  government subsidies, 141–42
  100th anniversary, 1–2
  invention of: aerodynamic control, 11–12; Chanute's contribution, 9–10; early gliders, 11; early tests, 12–13; first successful flight, 15; glider experiments, 14; Langley's contribution, 11; Lilienthal's contribution, 10–11; 100th anniversary celebration, 15; post-1903 designs, 15; propeller design, 14–15; propulsion problems, 14; Wright brothers, 91–96
  in linked formation, 169

retractable-gear monoplane, 41
U.S. production estimates 1939–41, 89
Aircraft carriers, 113
Aircraft design, 6; *see also* World War II aircraft ranking
  engine design, 141–42
  factor in World War II outcome, 139–41
  strategic effect, 114–15
  twin-engine *vs.* four-engine, 140–41
Aircraft engine development, 141–43
Aircraft engines, 141–42
Aircraft production
  pre–World War II, 29
  U.S. estimates 1939–41, 89
Air crew physiology, 46–47
Air doctrine
  in Germany post-1918, 84
  on precision bombing, 58
  recognition of, 38–39
  von Karman's projections, 163
Air Expeditionary Force, 34
Air Force Combat Command, 30
Air Force Logistics Command, 187
Air forces
  questions of function and control, 28–30
  theater-to-theater transfers, 113
Air Force Science Advisory Board, 45
Air Force Systems Command, 187, 189
Air Materiel Command, 186
Air-mindedness, 57
Air National Guard, 177
"Air News," 85
Air organization
  Army Regulation 95-5, 30
  Arnold's efforts after 1941, 32
  creation of Army Air Forces, 30
  creation of combined Chiefs of Staff 1941, 29
  critical stage 1939–41, 29
  and democratic process, 34–35
  independent Pacific command in 1944, 31
  interwar political problems, 27–29
  post–Cold War, 34–35
  pre–World War II buildup, 29–30
  as problem, 27
  Reorganization Act of 1986, 34
  reorganization in 1958, 33
  reorganization of 1942, 30–31
  and rise of Nazi Germany, 29
  tested in Korea, 33
  under Truman and Eisenhower, 32–33
  unified command concept, 31–32
  for Vietnam, 33–34
Air power
  and aircraft design, 114–15
  of Allies in World War II, 5

206 *Winged Crusade*

Index entries:

Independent Force, 56
intelligence assessments, 107
long-range escort problem, 60
losses over Germany, 112
Spitfires, 96
in World War I, 56
Royal Air Force aircraft
  AVRO Lancaster, 124–26
  Consolidated PBY, 131–32
  de Havilland Mosquito, 130–31
  Hawker Hurricane, 119–21
  Supermarine Spitfire, 121–22
Royal Air Force Bomber Command, 62
  AVRO Lancaster, 124–26
  in Battle of Britain, 112
  limited raids in Europe 1942, 60
  night flying in 1944, 63
  withdrawal of AVRO Manchester, 125
Royal Air Force Fighter Command, German assessment of, 93
Royal Navy
  Consolidated PB1 patrol aircraft, 131–32
  flight convoys, 111
  losses in Battle of Crete, 119
Ruchonnet, 16
Russia
  aircraft expenditures before 1914, 22
  pre-1914 aircraft development, 20
  secret training for German military, 83–84
Saar Valley steel works, 56
Salerno landing, 106
Salter, Roebrt, 185
SA-2 missiles, 73, 155–56, 159–60
Santos-Dumont, Alberto, 3, 15, 16, 18
Satellite and Missile Observation System, 185
Saudi Arabia, 75
Sayed Slim Kalay, 78
Scharnhorst sinking, 112
Schmid, Josef "Beppo," 89, 93
Schneider Trophy seaplane, 120
Schnorkel submarine, 74
Schriever, Bernard A., 2, 7, 8, 183–89
  achievements, 186
  and anti-satellite program, 187–88
  head of Air Force R&D Command, 186
  head of Air Force systems Command, 187
  head of Western Development Division, 185
  during Kennedy administration, 187
  legacy, 189
  lobbying for missiles/space program, 186
  and manned space flight, 188
  on space superiority, 183–84
Schwarzkopf, H. Norman, 74–75
Schweinfurt raid, 62, 103, 116–17, 135
Scientific American Trophy, 16
Scientific management, 59
Scud missiles, 75
Sea Hurricane Spitfire, 120–21
Sea war *versus* air war, 113

2nd Air Fleet, Germany, 95, 97–99, 107
2nd Bombardment Division, 116–17
Secretary of Defense, 33
Secretary of the Air Force, 33
Seguin, A., 17
Seifert, Frank, 40
Selective engagement strategies, 34
Serbia, 76
Seventh Air Force, 33
Shell Oil Company, 49–51
Sherman, Forrest, 33
Sherman, William C., 58
Short-range navigation radar, 69, 73
Sicily campaign of 1943, 97–99
Sicily landings, 106
Siebel, Franz, 90
Signal Corps of the *Luftwaffe*, 97, 102–3, 107
Signal Corps (U.S.), 40
Signals Branch of the Luftwaffe, 92
Sikorsky, Igor, 16–17
Single Integrated Operational Plan, 71
Skyraider, 176–77
Skytrain, 132
Smart, Jacob E., 29
Sound barrier, 24, 166
South Vietnam, 151
South Vietnamese insurgents, 72
Southwest Pacific Theater, 32, 113
Soviet aircraft
  Illushyn Il 2 Shturmovik, 118, 135–36
  Yakovlev Yak 1-9, 130
Soviet aircraft production, 93, 130
Soviet expansionism, 167
Soviet Union
  aid to North Vietnam, 72, 153
  aid to Vietnam, 6
  air force expansion, 73–74
  Berlin blockade, 168
  ICBMs, 71, 178
  jet engine development, 168–69
  launch of *Sputnik,* 186
  planning and executing air strikes against, 68
  planning for war with, 70–71
  stockpile of atomic weapons, 170
Spaatz, Carl Andrew, 27, 44, 62, 64, 93–94
Space age, 183
Space and Missile Systems Organization, 187–88
Space as vital military deterrent, 7
Space flights, 7
Space power
  anti-satellite program, 187–88
  ICBM development, 184–85
  Kennedy administration, 187
  manned flight, 188
  military satellite system, 185
  National Aeronautics and Space Administration founded, 183
  peaceful purposes policy, 186
  Schriever's achievements, 186–87